SpringerBriefs in Applied Sciences and Technology

Computational Mechanics

T0092294

Series Editors

Andreas Öchsner
Holm Altenbach
Lucas F. M. da Silva

For further volumes:
http://www.springer.com/series/8886

Mohammed Rafiq Abdul Kadir

Computational Biomechanics of the Hip Joint

 Springer

Mohammed Rafiq Abdul Kadir
Faculty of Health Science and Biomedical
   Engineering
Department of Biomechanics and
   Biomedical Materials
Universiti Teknologi Malaysia
Johor
Malaysia

ISSN 2191-5342   ISSN 2191-5350   (electronic)
ISBN 978-3-642-38776-0 ISBN 978-3-642-38777-7   (eBook)
DOI 10.1007/978-3-642-38777-7
Springer Heidelberg New York Dordrecht London

Library of Congress Control Number: 2013941344

Springer is part of Springer Science+Business Media (www.springer.com)

Preface

The number of people undergoing hip joint replacement surgery has increased over the past decades. In the UK alone, more than 60,000 total hip arthroplasties (THA) are performed annually, 15 % of which are performed in the younger age group (less than 57 years old) (Tennent and Goddard 2000). Most hip replacements are performed on patients suffering from osteoarthritis, a joint disease associated with the wearing away of the cartilage covering the bone ends. Other degenerative hip disorders that could require THA include rheumatoid arthritis and avascular necrosis. The primary aim of the replacement surgery is to relieve pain and regain mobility. Pioneered in 1962 by the renowned English surgeon, Sir John Charnley, the development of orthopaedic implants used in hip arthroplasty has improved steadily, making it one of the most successful surgical procedures. However, with the increase in the number of hip replacements performed, the scope and frequency of complications appear to be increasing. Complications such as stress-shielding, osteolysis and aseptic loosening remain some of the major problems in hip arthroplasty [1].

There are mainly two types of hip arthroplasty in use today—cemented and cementless. Hip prostheses with the use of cement are the most commonly used but the cementless techniques are gaining popularity. Fixation of these femoral components is a major concern because bone growth could only be achieved on stable implants (Pilliar 1991; Simmons et al. 1999). Failure to achieve a strong fixation will result in the formation of fibrous tissue layer at the bone-implant interface and the eventual loosening of the implant [2]. PMMA is used in the cemented type prostheses to provide strong primary fixation. However, cement debris can cause complications such as inflammation and bone lysis. One of the solutions to this problem is to abolish the use of cement, thus the cementless femoral component. However, without the cement these implants could not achieve initial fixation, unless the design is modified so that proper and adequate stability could be achieved. The design of femoral prostheses, together with the surgical techniques of implantation, continues to receive much attention in the hip biomechanics community.

This monograph will concentrate on the cementless hip replacement because instability is a cause of concern for this type of implant, more so than the cemented one. In this monograph, Finite Element Analyses (FEA) is used to investigate the issues of stability by calculating, through the use of a specially written computer code, the relative motion at the bone-implant interface. The quality of results for hip joint replacement depends on various factors, several of which such as the design, the surgical error and bone quality will be analysed in this monograph.

Malaysia, 2012   Mohammed Rafiq Abdul Kadir

References

1. Macdonald DA (1998) Mini symposium: Total hip replacement—(i) Risks versus rewards of total hip replacement. Curr Orthopaed 12 (4):229–231
2. Pilliar RM, Lee JM, Maniatopoulos C (1986) Observations on the effect of movement on bone ingrowth into porous-surfaced implants. Clin Orthop Relat Res (208):108–113

Contents

Notations

ABG	Anatomique benoist giraud
AML	Anatomic medullary locking
AVN	Avascular necrosis
BMD	Bone mineral density
CAD	Computer aided design
CT	Computed tomography
CoCr	Cobalt chromium
E	Young's modulus
FE	Finite element
GPa	Giga pascal
HA	Hydroxyapatite
inc	Inch
kN	Kilo newton
mm	Milimeter
MPa	Mega pascal
N	Newton
Nm	Newton meter
OA	Osteoarthritis
OP	Osteoporosis
SEM	Scanning electron microscopy
stl	Stereolithographic
TCP	Tricalcium phosphate
THA	Total hip arthroplasty
TiAl	Titanium alloy
VHP	Visible human project
WHO	World health organisation
x, y, z	Cartesian coordinates
2D	Two-dimensional
3D	Three-dimensional
μm	Micrometer
%	Percentage
°	Degree

Chapter 1
Introduction

Abstract This chapter introduces the hip joint and hip disorders that may lead to total replacement of the joint. The internal morphology of the proximal femur, which can be categorised into one of three types, is of primary importance as it will affect the type of implant most suitable for the treatment. Osteoarthritis is one of the most common reasons for hip replacement where in certain cases can severely limit patient mobility and reduce quality of life. Implants for total hip arthroplasty (THA) can be categorised into one of two types, cemented and cementless. Whilst cemented has been regarded as the gold standard, the cementless coutnerpart is gaining popularity. However, the issue of primary stability has to be addressed and tackled if the cementless approach was to be widely accepted. Finite element method will be used to analyse the stability of the cementless implants for hip arthroplasty.

Keywords Hip joint • Hip disorders • Arthroplasty • Primary stability • Finite element method

1.1 The Anatomy and Physiology of the Hip Joint

The hip bone consists of three fused bone parts, the ilium, ischium and pubis. There is a large cup-shaped articular cavity, called the acetabulum, acted as the socket for articulation with the head of the femur. The ilium is the superior broad and the ischium is the lowest and strongest portion of the bone. The pubis which extends medialward and downward from the acetabulum forms the front part of the pelvis.

The femur is the longest human bone and can be divided into three parts; proximal, middle and distal. The proximal part consists of a head, a neck, a greater trochanter and a lesser trochanter. The hemispherical head forms a ball-and-socket joint with the acetabulum via the femoral head ligament and by strong surrounding ligaments. The neck of the femur connects the shaft and head at an average normal angle of 125°. The Greater Trochanter provides leverage to the muscles that rotate

M. R. Abdul Kadir, *Computational Biomechanics of the Hip Joint*, SpringerBriefs in
Computational Mechanics, DOI: 10.1007/978-3-642-38777-7_1,
© The Author(s) 2014

the thigh on its axis whilst the Lesser Trochanter gives the insertion to the tendon of Psoas Major and Illiacus. The internal structure of the femur shows arcs of trabeculae that are efficiently arranged to transmit pressure and resist stress.

The femoral shaft is almost cylindrical in form and somewhat convex forward. The linea aspera—a prominent longitudinal ridge—strengthened the posterior shaft and provides attachment for several adductor muscles. The distal part of the shaft is larger than the proximal and consists of two condyles which formed the upper half of the knee joint. The medial condyle is more prominent than the lateral condyle and are separated from one another by the patellar surface, a smooth shallow articular depression.

1.1.1 Femoral Morphology Classification

The proximal femur has considerable variation in morphologic features, and the density and structure changes with age. The quality and morphology of the proximal femur has been described and radiographically categorised into types A, B and C, in an attempt to guide surgical technique and implant selection during total hip arthroplasty [1, 2]. Type A has thick cortices and a narrow diaphyseal canal which produce a funnel or champagne-fluted shape to the proximal femur. Type C has virtually lost the medial and posterior cortices, with the anterior cortices may also be dramatically thinned. The intramedullary canal is usually very wide resulting in a cylindrical or stovepipe shape. Type B has an intermediate funnel shape between type A and type C (Fig. 1.1).

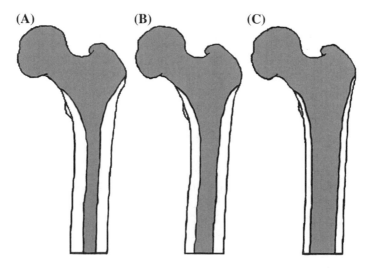

Fig. 1.1 Schematic representation of the proximal femur showing the three different internal morphologies—type A, B and C

Fig. 1.2 Diagrammatic representation of the canal flare index

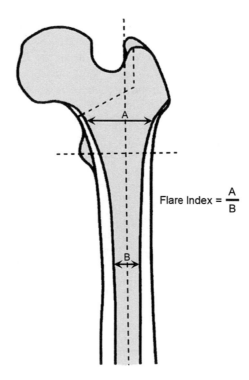

Flare Index = $\dfrac{A}{B}$

The three types of femoral morphology were expressed by a single geometric parameter called the canal flare index, defined as the ratio of the intracortical width of the femur at a point 20 mm proximal to the lesser trochanter and at the canal isthmus (Fig. 1.2) [2]. From a study of two hundred femora with age ranging from 22 to 95 years, Noble et al. [2] found that canal flare indices of less than 3.0 described stovepipe canals (type C), 3.0–4.7 normal canals (type B), and 4.7–6.5, canals with a champagnefluted appearance (type A).

1.1.2 Muscles of the Femur

There are many muscles that are involved in the motion of the hip and they can be categorised into several groups depending on the type of movements; flexion-extension (sagittal plane), abduction–adduction (frontal plane) and rotation (transverse plane. Real movement is usually a combination of these pure movements.

The Psoas and the Iliacus are the most powerful flexor muscles with the longest range. It can also produce adduction/abduction or lateral/medial rotation as accessory movements. The tensor fascia latae, which is one of the muscles in the gluteal region, is also a fairly powerful flexor. The gluteus maximus is the largest and strongest muscle of the body with two other glutei muscles, medius and minimus, assisted its function as extensors. These muscles are also lateral rotators.

The main abductor muscle is the Gluteus medius, which can also produces abduction with its highest and superficial fibres. The Gluteus minimus, is slightly smaller in size and produce a force one-third that of the Gluteus medius. The Piriformis and the tensor fascia latae are also muscles involved in abduction. The Adductor magnus which arises from the pubis, the ischium, and the tuberosity of the ischium is the most powerful of the adductors. The anterior–superior portion is often described as a separate muscle, the Adductor minimus. The large bulk of the gluteus maximus muscle is also adductors.

There are numerous lateral rotator muscles in the hip, some of which are the piriformis and the gluteus maximus. The medial rotators are less numerous than the lateral rotators. The three main medial rotator muscles are the tensor fascia latae, the gluteus minimus and the gluteus medius.

1.2 Hip Joint Diseases and Skeletal Disorders

Osteoarthritis is a degenerative condition associated with the wearing away of the cartilage covering the bone-ends. Cartilage is used to absorb the stresses put on a joint, and protects the bones from damage. Osteoarthritis occurs when the cartilage deteriorates, either due to age or injury. With age, human joints slowly lose the ability to regenerate and repair the cartilage. Bone growths develop as the cartilage degenerates, and the bones that make up the joint rub together, causing pain and restriction of movement. It is a progressive disease that can affect any or all of the joints in the human body, with the weight-bearing joints such as the hips and the knees being more susceptible than the others. Currently, no cure exists for osteoarthritis. Treatment options centre on prevention, if possible, and control of the disease. At its worst, the disease can cause constant pain and severely reduced mobility.

Rheumatoid arthritis is another form of arthritis that is chronic and affects many different joints. It begins with an inflammation and thickening of the synovial membrane, which causes pain and swelling, followed by bone and cartilage degeneration and disfigurement. The disease is considered to be an autoimmune condition that is acquired, and in which genetic factors appear to play a role. It appears more frequently in older people, with more women being affected than men.

Avascular Necrosis (AVN) is a skeletal disease that occurs when bone tissues die off, resulting in the collapse of the bone. It is caused by the loss of blood to the bone. Hips are one of the areas most commonly affected by the disease. Common names for this disease include osteonecrosis, aseptic necrosis and ischemic bone necrosis. The disease can be caused by excessive use of drugs and alcohol or by injury. A joint that has been injured through fracture or dislocation has an increase risk of AVN because blood vessels may be damaged, and blood circulation to the bone is disrupted resulting in trauma-related AVN. Arthritis and AVN affect both men and women at higher rates as they age. The majority of cases involve people between the ages of 30 and 50 years. However, it can affect people of all ages.

Osteoporosis is a major skeletal disorder in which nutrition plays a role and can be prevented and treated. It is characterised by a significant loss of cancellous bone stock and structural deterioration of bone tissue, causing it to become fragile and more likely to fracture. If left untreated, the disease can progress without symptoms until the bones became so weak that a sudden strain, bump, or fall causes a fracture. Fractures occur typically in the hip, spine and wrist. Women are more likely to suffer from osteoporosis than men. Osteoporosis is not a cause for joint replacement, but if a replacement is required due to fracture for example, then it affects the decision made by the surgeon because of the structural deterioration of the bone.

1.3 Hip Arthroplasty

People suffering from severe hip joint diseases, may choose to have their joint replaced. Total hip arthroplasty is made up of two major parts, the acetabular component and the femoral component. The acetabular component (socket portion) is a hemispherical cup that replaces the acetabulum. It comprises of a shell with an inner socket liner that acts like a bearing. The femoral component (stem portion) replaces the femoral head (Fig. 1.3).

Whilst the use of hip implants has helped patients suffering from skeletal diseases or injuries to relieve pain and gain mobility, its use in human body is not without its own shortcomings. As with all major surgical procedures, complications can occur. Failures of these implants have been reported and many studies have been conducted to identify the reasons behind the failures. Some of the most common complications following hip replacement are as follows [3]:

1. Infection

Infection can be a very serious complication following an artificial joint surgery. Some infections may show up very early—even before the patient leaves the hospital. Others may not become apparent for months, or even years, after the operation. Infection can spread into the artificial joint from other infected areas. Prevention of infection involves a combination of many factors such as ensuring clean air theatres, and the use of antibiotic-containing cement in the case of cemented hip arthroplasty.

2. Dislocation

Just like a natural hip, an artificial hip can dislocate, where the ball comes out of the socket. There is a greater risk just after surgery, before the tissues have healed around the new joint, but there is always a risk. There are certain activities and positions which the patient must avoid in order to prevent hip dislocation. A hip that dislocates more than once may have to be revised to make it more stable.

Fig. 1.3 Major components of a hip joint (**a**) and major components of a THA (**b**)

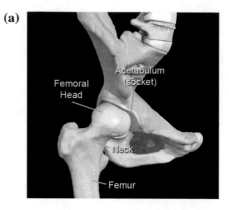

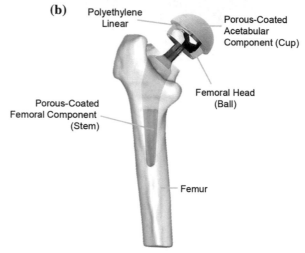

3.   Stress-Shielding

Proximal bone loss due to stress-shielding is also a cause of concern because it could lead to loosening. It is an adverse bone-remodelling phenomenon where bone is resorbed in areas where it is not loaded to physiological levels. This unwanted bone remodelling could cause failure of arthroplasty due to the weakening and deterioration of the proximal bone mass.

4.   Aseptic Loosening

The major reason that artificial joints eventually fail continues to be through a process of loosening where the implant or cement meets the bone. A loose hip is a problem because it causes pain. Once the pain becomes unbearable, another operation will probably be required to revise the hip. There are many factors that cause the loosening of hip prostheses. One of the common causes of aseptic loosening is an inflammatory reaction to particles or wear debris [4]. Wear of polyethylene, metal and bone cement produces debris particles that induce bone resorption (osteolysis) and implant loosening. These causes further increase in wear debris and a vicious cycle of failure begins. This cycle of wear subsequently causes implant migration, periprosthetic fractures, dislocation and pain [3].

Another common mechanism of loosening is micromotion of implants that did not achieve adequate initial fixation. Without good fixation, bone cannot grow into the implant in the desired way and so cannot attach onto the surface of the implant. This is especially important for implants with relatively smooth surfaces, where fibrous tissue formation would take precedence over bone formation.

1.3.1 Type of Hip Implant

There are two major types of artificial hip replacements: cemented and uncemented. Both are widely used. The superiority of one compared to the other remains a topic of debate. The choice made by orthopaedic surgeons is usually based on the patient's age, type of disease, anatomy, lifestyle, and the surgeon's experience. The surgeon's criteria of a suitable hip implant include:

- Shape and design for optimum fit within the patient's anatomy.
- Shape and design which allow for optimum surgical technique.
- The implant's clinical history of stability—called fixation.
- The exterior coating of the stem, which contributes to optimum fixation.
- The type of surgery—primary, revision or fracture.

Cemented prostheses are the oldest form of hip arthroplasty. They are held in place by a type of epoxy cement that attaches the stem to the bone. Good clinical results at 10 years are achievable through this method [5]. There are, however, complications arising from this conventional method of fixing artificial joints, mainly from the use of the cement itself. PMMA degrades over time and the built-up of cement debris causes inflammation of the surrounding bone and loosening at the interface [6]. Revision surgery, if required, will also be a problem due to the significant removal of bulk cancellous bone during the first surgery. Despite these complications cemented techniques are still widely used depending on the condition of the patients. It has been suggested for use in the older patients above 65 years [7], or on patients with osteoporotic bone condition or poor bone stock to ensure strong primary fixation [8]. However, Harris [9] strongly believed that cemented techniques not only worked well with the elderly, but also for younger patients.

Early total hip arthroplasty using the cementing technique in the late 1960s and early 1970s showed an incidence of femoral component loosening of 30–40 % at 10 years [10, 11]. These disappointing results led to revised cementing techniques [12], and the use of bioactive cement to encourage rapid bone ingrowth [13]. Some authors also suggested limiting the cement to the proximal area only (sometimes called the hybrid option) [14].

The cementless alternative is used to eliminate the problems associated with the use of cement. The diameter of the stem is generally larger than its cemented counterpart in order to fill the canal, and the surface is usually roughened with particles through gritblasting or plasma-spraying. Roughened surfaces are not recommended in cemented stems; they are usually polished in order to reduce cement wear debris [15]. Some cementless designs also have macro-features at the proximal part such as a porous coating or fibre meshing. These features are beneficial in cementless implants because they provide a medium for biological attachment that secures the implant onto its place. The roughened surface encourages bone attachment and the macro-features allow new bone to attach and grow into, creating a strong bond between the patient's own natural tissue and the implant [16].

The earlier generation of cementless stems performed less favourably than the cemented designs at that time [17]. However, most cementless hip stems currently in use have survival rate similar to their cemented counterpart [18]. One study reported that clinical results for cementless designs were comparable to cemented stems in terms of disease specific, patient specific, global health, or functional capacity [19]. Another study showed that statistically, there were no significant differences in clinical and functional outcome between cementless and cemented femoral components in the first 12 months post surgery [20].

Besides the promising results of cementless stems, old complications associated with the cemented technique are still not entirely eliminated. Stress-shielding, for example, still occurs in hip replaced with cementless stems [21]. Osteolysis also occurs in cementless arthroplasty due to wear debris. It was previously thought that migration of cement debris along the bone-cement interface was the cause of osteolysis, due to the detection of particles of PMMA of various sizes found in failed interface membranes. However, a study by Boss et al. [22] showed that the release and deposition of wear products, of whatever nature, were apparently responsible for osteolysis.

One of the major problems associated with cementless stems is thigh pain. This has been attributed to the lack of fixation, i.e. the occurrence of interface motion between the stem and the femoral shaft [23]. This was confirmed by Cruz-Pardos & Garcia-Cimbrelo [24] who found a high correlation of thigh pain with unstable fixation. Whiteside (1989) reported a similar observation; that a tight distal fit correlated with a lower incidence of thigh pain. Pain was significantly more likely to occur in those patients with a loose distal fit (20 out of 38, i.e. 53 %) than in those with a tight distal fit (2 out of 67, i.e. 3 %).

Debate also exists on the suitability of cementless stems in bone with stovepipe characteristic (type C bone). Some authors have recommended the use of cementless femoral fixation for patients with type A and type B only [1, 18, 25]. Healy [26] however, reported that cementless stems were not just reliable with younger patients who have good bone stock, but were also reliable for elderly patients (above 75 years of age) with type C bone. Reitman et al. [27] also reported that cementless stems could be used in bones with a stovepipe characteristic.

Aseptic loosening is a major concern in cementless hip stems, and it is a failure associated with the lack of stability. It can be divided into 2 categories [28]. The 1st category is a late failure due to osteolysis which can be explained by excessive wear particles that migrate along the implant-bone interface and result in formation of the activated interfacial membranes. Early failure, on the other hand, can be explained by the lack of initial implant stability or catastrophic failure of metal or polyethylene materials. Aseptic loosening is still a major problem in cementless design. This could be seen in a recent follow-up report of a carbon fibre-reinforced composite stems, where 92 % of the stems were revised due to aseptic loosening at 6 years [29]. The authors blamed the high rate of loosening on deficient primary stability due to the bad design.

1.4 Primary Stability in Cementless Stems

Achieving good primary fixation is of crucial importance in cementless hip arthroplasty to ensure good short and long term results. One of the major direct consequences of a lack of stability is the eventual loosening of the prosthesis. This stability, or the lack of it, is commonly measured as the amount of relative motion at the interface between the bone and the stem under load. Large interfacial relative movements reduce the chance that bone will osseointegrate with the implant, and may cause the formation of a fibrous tissue layer around the prosthesis. The shear strength at the interface will reduce significantly as well as the ability to transfer load to the surrounding bone. This will further encourage the formation and thickening of the fibrous tissue, which eventually leads to loosening and failure of the arthroplasty.

Achieving good initial stability is also important because if this is attained at surgery, immediate post-operative full weight bearing is acceptable—i.e. protected weight bearing is unnecessary [30]. Another similar study also showed that when solid initial fixation was obtained intraoperatively and radiographically, bone ingrowth reliably occurred whether or not a partial or full weight-bearing postoperative protocol was followed [31].

Implant fixation by bone ingrowth is one of the aims of hip arthroplasty. The term bone ingrowth or osseointegration refers mainly to bone formation within a porous surface structure of an implant, but it can also refer in general to any bone formation into the irregular depths of non-smooth surfaces or an intimate implant surface-to-bone contact. Clinical success of fixation by bone ingrowth depends on a stable implant-bone interface. This primary stability can vary depending on implant design variables (geometry, means of additional fixation, stiffness mismatch), implantation technology variables (accuracy of tools for rasping, broaching, reaming, drilling, sawing), surgical technique variables (accuracy of use of the implantation technology), and patient variables (bone quality, bone defects) [32].

Cementless hip stems, though eliminate the problems associated with the use of cements, lack the very thing that the cement provides—strong initial fixation. Several in vitro experimental studies found larger micro-movement for the cementless stems compared to the cemented stems. Burke et al. [33] found that the micromotion was always roughly about twice that of the cementless design at low loads and up to 3.5 times at higher loads. In a comparative study between a cemented stem and four cementless stems, Schneider et al. [34] found that all cementless stems tested in vitro produced more micromotion than the cemented. Another comparative study also showed similar results [35].

Clinically, aseptic loosening is one of the major concerns in cementless hip arthroplasty. In a two-year follow-up report of the Bichat III cementless stem [36], 16 % from a total of 203 hips were revised due to persistent thigh pain. All the revised stems were found to be loose. From the unrevised stems 60 % of them also complained of pain, and their symptoms were attributed to micromotion of the implant relative to the endosteal bone. Dickob and Martini [37] reported that the cementless PM smooth prosthesis had 15 % of its femoral components revised for loosening at 6 years. From the unrevised hip replacements, 73 % were not completely free from pain. Menon and McCreath [38] reported that the cementless Freeman stems were unsatisfactory at mid-term results. 30 % were revised, 16 % of which were due to aseptic loosening at an average follow-up period of 4–5 years. There was a high incidence of thigh pain (40 %) and an average femoral subsidence was 5.4 mm. A study on the smooth surface Moore stem showed that thigh pain on weight bearing was the main problem and was due to loosening of the stem [39]. The follow up of cementless APR-I stems at 6.7 years showed that 70 % of the stems had progressive loss of fixation [40].

Radiographic signs that were correlated with instability have been reported. Kobayashi et al. [41] concluded in his paper that predictors of aseptic loosening was either migration of ≥ 2 mm at 2 years or the presence of radiolucent lines of 2 mm occupying 1/3 of any zone. Vresilovic et al. [42] reported that radiolucent lines covering greater than 50 % of the bone-prosthesis interface was one of the radiographic signs correlated with instability. Khalily and Whiteside [43] also reported the use of radiolucent lines as a predictor of future need for revision. However, there were cases where the implant migrated significantly in the first year and then stabilised—usually termed late stabilisation [44]. No complications were found later on for these late-stabilised implants. Lautiainen et al. [45] argued that subsidence was not a definitive hallmark of loosening between 2 and 5 years follow-up. In a 4–8 years follow-up study of Harris-Galante cementless femoral stem [46], 22 % of the stems were found to have migrated more than 2 mm but later stabilised.

In order to improve its stability, cementless stems are usually designed with rough surfaces or surfaces with microgeometry. Mont and Hungerford [47] reported that proximal microfeature of any kind improved the success of cementless stems. These include the interconnected multilayered microporous beads and mesh coatings, to other various microporous coated components in use today such as plasma sprayed surfaces and variegated textured surfaces. Chen et al. [48] reported a 5-year follow-up study of porous-coated cementless stems where 92 % of the stems showed evidence of bone ingrowth into the porous coating. A strong fixation can also be achieved with multilayered honeycomb mesh structure etched onto the stem's surface where this structure provides the anchoring effect through mechanical interlocking [49]. Circumferential coating in cementless stems also has the added advantage of sealing the diaphysis from wear debris, thus preventing osteolysis in the distal region [50]. A comparative study between hip stems with and without circumferential coating showed that the survival rate of hip stems with the circumferential coating was significantly better [40, 51, 52].

It has also been reported that these microtextures are able to influence bone-cell behaviour [53]. Apart from encouraging the formation of osteoblastlike cells [54], microtextured and microgrooves also helped in anchoring the bone because it has been proven that these features can determine the alignment of cells and cellular extensions [53, 55].

Biological fixation is achievable by the use of bioactive materials such as hydroxyapatite [56–59], other calcium-phosphate based materials [60–62], calcium ion (Ca^{2+}) [63] or glass and glass–ceramics [64]. The effect can also be further enhanced by covering the surface of these materials with synthetic stimulants [65, 66]. All these bioactive materials have been shown to improve the amount of bone apposition to the implant's surface. The attachment of bone does not just show intimate contact, but also exhibits a chemical bond to the surface. Resorption or dissolution will take place, over time, where new bone will replace the resorbed materials [67, 68]. The rate of resorption is therefore crucial to the success of these biomaterials. Rapid resorption could lead to disintegration of the coating with loss of bonding strength and mechanical fixation.

There are, however, published reports that showed the uncertain short-term effect of bioactive coatings. Ciccotti et al. [69] found that there were no differences between implants coated with and without HA coatings at 6 and 12 weeks in terms of plain radiographic evaluations. No instability was detected and bone ingrowth was similar in the two groups. Rothman et al. [70] also reported that there were no significant clinical or radiographic advantages between HA-coated and non HA-coated implants 2 years after surgery. The same conclusion was also arrived at by Johnston et al. [71] when comparing standard AML stem with the one coated with tricalcium phosphate (TCP). The two groups of 46 subjects showed no significant differences in terms of bone ingrowth, the degree of hypertrophy/calcar atrophy at 6, 12 and 24 months. Jinno et al. [63] reported greater bone apposition in calcium ion coated prostheses as compared to the non-coated, but this was only significantly different at 1 month. They concluded that calcium ion was beneficial only for early fixation unless if the dissolution rate could be controlled.

Longer follow-up study also showed that there were no significant advantageous of using HA coating. A study of IPS cementless stems with and without HA coating showed that clinical and radiographic results were similar at a mean duration of 6.6 years [72]. No aseptic loosening was found in both groups and the bone remodelling patterns, including calcar atrophy, were similar. In a study comparing between HA and non-HA porous coated Mallory-Head stems, no significant differences were found in terms of Harris hip scores and femoral stem survivorship at a minimum of 3 years [73]. Kang et al. [74] reported that there were no statistical clinical or radiographical differences between the HA-coated APR-II stems and the noncoated at 4 years. But there were significant differences between proximally porous coated APR-II and the non-porous coated, with the latter having 14 % rate of osteolysis and 0 % for the former.

In a study comparing between implants with or without HA coating on grooved specimens, Hayashi et al. [75] reported that geometry played a more important role than HA coating alone. Specimens with 1 mm grooves HA-coated showed higher percentage of bone ingrowth in 4 weeks, but showed minimal improvements in the following 8 and 12 weeks. By 8 weeks, the uncoated control specimens showed similar percentage of bone ingrowth. However, specimens with 2 mm grooves HA coated showed similar ratio of bone ingrowth to the uncoated 1 mm grooves. Furthermore, specimens with porous-coated beads showed lower attachment strength at 4 weeks compared to HA coated grooves, but significantly stronger attachment strength at 12 weeks. It was concluded that the percentage of bone ingrowth increased rapidly with HA coating at a specific type of geometry, with certain types of geometry, such as the beads, proving to be better than HA coating alone.

The threshold value of relative micromotion, above which fibrous tissue layer forms, has been studied in both animals and human. In a review of dental implants in animals, the threshold micromotion value was found between 50 and 150 μm [76]. The tolerated micromotion threshold varies according to surface state and/ or implant's design. Femoral implants in animal, on the other hand, gave a slightly lower threshold value of 30 μm. A similar range of threshold values were also reported for orthopaedic implants in human. A micromotion study on eleven cemented femoral specimens retrieved at autopsy found a maximum axial micromotion of 40 μm [77]. Histologic investigation showed intimate oseeointegration at the interface with only rare intervening fibrous tissue. The same magnitude of micromotion was found in cementless femoral components with bone ingrowth at the porous coating and a higher micromotion of 150 μm was found on areas of failed bone ingrowth [78]. Another micromotion study comparing the AML and the Mallory Head prosthesis with surface bone ingrowth showed a micromotion of 80 μm or less [16]. It can be concluded from these experiments that the threshold value of micromotion for osseointegration is between 30 and 150 μm.

Computer models simulating the effect of fibrous tissue layer have also been reported in the literature. In a study comparing the effect of various fibrous tissue layer thicknesses on micromotion with a threshold value for bone ingrowth of 200 μm, a thickness of 58 μm was found to be sufficient to increase the micromotion beyond the targeted threshold value [79]. It was also reported from a similar study [80], that the viable region for bone ingrowth was completely eliminated throughout the stem surface for a fibrous tissue layer of thickness 300 μm. Others have used a mathematical model which incorporates the effect of interface debonding and relative motions on bone resorption [81]. Three simplified models of orthopaedic implants were analysed, and the results showed reasonable qualitative agreement with resorption patterns found in clinical studies.

1.4.1 The Use of Finite Element Analysis in Orthopaedic Biomechanics

Finite element (FE) models of joint anatomy can help surgeons understand trauma from repetitive stress, degenerative diseases such as osteoarthritis, and acute injuries. Finite element models of prosthetic joint implants can provide surgeons and biomechanical engineers with the analytical tools to improve the life-span of implants and improve the clinical outcomes of total hip replacement surgeries [82]. These models are created from many small "elements" of triangular or rectangular shapes. When these FE models are loaded with proper boundary conditions, their responses are obtained by solving a set of simultaneous equations that represent the behaviour of the model under load.

Finite element analyses have been widely used in the study of hip joint arthroplasty and in particular the study of hip stem stability. Finite element analysis is used to complement the experimental work on micromotion and could even become a useful tool to assess the suitability of implants before surgery [83, 84]. One advantage of using FE methods in analysing the stability of hip stems is that it is a non-destructive assessment that can measure the distribution of micromotion along the entire surface of the stem. *In-vitro* experiments, on the other hand, could only measure micromotion at certain points, and the drilling of holes during specimen preparation could damage and weaken the surrounding bone, thus overestimating the relative motion. Finite element analysis gives a clearer picture of the stem being analysed and could pointed out ways of improving the stability further. Finite elements have been shown to predict experimental findings and long-term failure mechanisms in orthopaedic surgery with excellent accuracy [85, 86].

With faster computers and more reliable software, computer simulation is becoming an important tool in orthopaedic research. Future research programmes will use computer simulation to reduce the reliance on animal experimentation, and to complement clinical trials [87].

References

1. Dorr LD, Faugere M-C, Mackel AM, Gruen TA, Bognar B, Malluche HH (1993) Structural and cellular assessment of bone quality of proximal femur. Bone 14(3):231–242
2. Noble PC, Alexander JW, Lindahl LJ, Yew DT, Granberry WM, Tullos HS (1988) The anatomic basis of femoral component design. Clin Orthop Relat Res 235:148–165
3. Macdonald DA (1998) Mini symposium: total hip replacement—(i) Risks versus rewards of total hip replacement. Curr Orthopaed 12(4):229–231
4. Bauer TW, Schils J (1999) The pathology of total joint arthroplasty—II. Mech Implant Fail Skeletal Radiol 28(9):483–497
5. Emery D, Britton A, Clarke H, Grover M (1997) The stanmore total hip arthroplasty: A 15- to 20-year follow-up study. J Arthroplasty 12(7):728–735
6. El Warrak AO, Olmstead ML, von Rechenberg B, Auer JA (2001) A review of aseptic loosening in total hip arthroplasty. Vet Comp Orthopaed 14(3):115–124
7. Dorr LD, Wan Z, Gruen T (1997) Functional results in total hip replacement in patients 65 years and older. Clin Orthop Relat Res 336:143–151
8. Haber D, Goodman SB (1998) Total hip arthroplasty in juvenile chronic arthritis: a consecutive series. J Arthroplasty 13(3):259–265
9. Harris WH (1997) Options for primary femoral fixation in total hip arthroplasty. Cemented stems for all. Clin Orthop Relat Res 344:118–123
10. Stauffer RN (1982) 10-year follow-up-study of total hip-replacement—with particular reference to roentgenographic loosening of the components. J Bone Joint Surg Am 64(7):983–990

11. Sutherland CJ, Wilde AH, Borden LS, Marks KE (1982) A 10-year follow-up of 100 consecutive Muller curved-stem total hip-replacement arthroplasties. J Bone Joint Surg Am 64(7):970–982
12. Rasquinha VJ, Dua V, Rodriguez JA, Ranawat CS (2003) Fifteen-year survivorship of a collarless, cemented, normalized femoral stem in primary hybrid total hip arthroplasty with a modified third-generation cement technique. J Arthroplasty 18(7 Suppl 1):86–94
13. Oonishi H (1991) The bone-biomaterial interface. Interfacial reactions to bioactive and non-bioactive bone cements. University of Toronto Press, Ontario
14. Monti L, Cristofolini L, Viceconti M (2001) Interface biomechanics of the Anca dual fit hip stem: an in vitro experimental study. Proc Inst Mech Eng H 215(6):555–564
15. Duffy GP, Muratoglu OK, Biggs SA, Larson SL, Lozynsky AJ, Harris WH (2001) A critical assessment of proximal macrotexturing on cemented femoral components. J Arthroplasty 16(8 Suppl 1):42–48
16. Whiteside LA, White SE, Engh CA, Head W (1993) Mechanical evaluation of cadaver retrieval specimens of cementless bone-ingrown total hip arthroplasty femoral components. J Arthroplasty 8(2):147–155
17. Kim YH, Oh SH, Kim JS (2003) Primary total hip arthroplasty with a second-generation cementless total hip prosthesis in patients younger than fifty years of age. J Bone Joint Surg Am 85-A (1):109–114
18. Bourne RB, Rorabeck CH (1998) A critical look at cementless stems. Taper designs and when to use alternatives. Clin Orthop Relat Res 355:212–223
19. Mulliken BD, Nayak N, Bourne RB, Rorabeck CH, Bullas R (1996) Early radiographic results comparing cemented and cementless total hip arthroplasty. J Arthroplasty 11(1):24–33
20. Zimmerma S, Hawkes WG, Hudson JI, Magaziner J, Hebel JR, Towheed T, Gardner J, Provenzano G, Kenzora JE (2002) Outcomes of surgical management of total HIP replacement in patients aged 65 years and older: Cemented versus cementless femoral components and lateral or anterolateral versus posterior anatomical approach. J Orthop Res 20(2):182–191
21. Nourbash PS, Paprosky WG (1998) Cementless femoral design concerns. Rationale for extensive porous coating. Clin Orthop Relat Res 355:189–199
22. Boss JH, Shajrawi I, Soudry M, Mendes DG (1990) Histological features of the interface membrane of failed isoelastic cementless prostheses. Int Orthop 14(4):399–403
23. Engh CA, Bobyn JD, Glassman AH (1987) Porous-coated hip replacement. The factors governing bone ingrowth, stress shielding, and clinical results. J Bone Joint Surg Br 69(1):45–55
24. Cruz-Pardos A, Garcia-Cimbrelo E (2001) The Harris-Galante total hip arthroplasty: a minimum 8-year follow-up study. J Arthroplasty 16(5):586–597
25. Barrack RL (1998) The adult hip. Pre-operative planning, 1st edn. Lippincott-Raven, USA
26. Healy WL (2002) Hip implant selection for total hip arthroplasty in elderly patients. Clin Orthop Relat Res 405:54–64
27. Reitman RD, Emerson R, Higgins L, Head W (2003) Thirteen year results of total hip arthroplasty using a tapered titanium femoral component inserted without cement in patients with type C bone. J Arthroplasty 18:116–121
28. Horikoshi M, Macaulay W, Booth RE, Crossett LS, Rubash HE (1994) Comparison of interface membranes obtained from failed cemented and cementless hip and knee prostheses. Clin Orthop Relat Res 309:69–87
29. Adam F, Hammer DS, Pfautsch S, Westermann K (2002) Early failure of a press-fit carbon fiber hip prosthesis with a smooth surface. J Arthroplasty 17(2):217–223
30. Chan YK, Chiu KY, Yip DK, Ng TP, Tang WM (2003) Full weight bearing after noncemented total hip replacement is compatible with satisfactory results. Int Orthop 27(2):94–97
31. Woolson ST, Adler NS (2002) The effect of partial or full weight bearing ambulation after cementless total hip arthroplasty. J Arthroplasty 17(7):820–825
32. Kienapfel H, Sprey C, Wilke A, Griss P (1999) Implant fixation by bone ingrowth. J Arthroplasty 14(3):355–368

33. Burke DW, O'Connor DO, Zalenski EB, Jasty M, Harris WH (1991) Micromotion of cemented and uncemented femoral components. J Bone Joint Surg Br 73(1):33–37
34. Schneider E, Kinast C, Eulenberger J, Wyder D, Eskilsson G, Perren SM (1989) A comparative study of the initial stability of cementless hip prostheses. Clin Orthop Relat Res 248:200–209
35. Schneider E, Eulenberger J, Steiner W, Wyder D, Friedman RJ, Perren SM (1989) Experimental method for the in vitro testing of the initial stability of cementless hip prostheses. J Biomech 22(6–7):735–744
36. Duparc J, Massin P (1992) Results of 203 total hip replacements using a smooth, cementless femoral component. J Bone Joint Surg Br 74(2):251–256
37. Dickob M, Martini T (1996) The cementless PM hip arthroplasty. Four-to-seven-year results. J Bone Joint Surg Br 78(2):195–199
38. Menon DK, McCreath SW (1999) 5- to 8-Year results of the freeman press-fit hip arthroplasty without HA coating: a clinicoradiologic study. J Arthroplasty 14(5):581–588
39. Phillips TW, Messieh SS (1988) Cementless hip replacement for arthritis. Problems with a smooth surface moore stem. J Bone Joint Surg Br 70(5):750–755
40. Dorr LD, Lewonowski K, Lucero M, Harris M, Wan Z (1997) Failure mechanisms of anatomic porous replacement I cementless total hip replacement. Clin Orthop Relat Res 334:157–167
41. Kobayashi A, Donnelly WJ, Scott G, Freeman MA (1997) Early radiological observations may predict the long-term survival of femoral hip prostheses. J Bone Joint Surg Br 79(4):583–589
42. Vresilovic EJ, Hozack WJ, Rothman RH (1994) Radiographic assessment of cementless femoral components: correlation with intraoperative mechanical stability. J Arthroplasty 9(2):137–141
43. Khalily C, Whiteside LA (1998) Predictive value of early radiographic findings in cementless total hip arthroplasty femoral components: an 8- to 12-year follow-up. J Arthroplasty 13(7):768–773
44. Kitamura S, Hasegawa Y, Iwasada S, Yamauchi K-i, Kawamoto K, Kanamono T, Iwata H (1999) Catastrophic failure of cementless total hip arthroplasty using a femoral component without surface coating. J Arthroplasty 14(8):918–924
45. Lautiainen IA, Joukainen J, Makela EA (1994) Clinical and roentgenographic results of cementless total hip arthroplasty. J Arthroplasty 9(6):653–660
46. Petersilge WJ, D'Lima DD, Walker RH, Colwell CW Jr (1997) Prospective study of 100 consecutive Harris-Galante porous total hip arthroplasties. 4- to 8-year follow-up study. J Arthroplasty 12(2):185–193
47. Mont MA, Hungerford DS (1997) Proximally coated ingrowth prostheses. A review. Clin Orthop Relat Res 344:139–149
48. Chen CH, Shih CH, Lin CC, Cheng CK (1998) Cementless Roy-Camille femoral component. Arch Orthop Trauma Surg 118(1–2):85–88
49. Kusakabe H, Sakamaki T, Nihei K, Oyama Y, Yanagimoto S, Ichimiya M, Kimura J, Toyama Y (2004) Osseointegration of a hydroxyapatite-coated multilayered mesh stem. Biomaterials 25(15):2957–2969
50. von Knoch M, Engh CA, Sychterz CJ, Engh CA, Willert H-G (2000) Migration of polyethylene wear debris in one type of uncemented femoral component with circumferential porous coating: an autopsy study of 5 femurs. J Arthroplasty 15(1):72–78
51. Dorr LD, Wan Z (1996) Comparative results of a distal modular sleeve, circumferential coating, and stiffness relief using the anatomic porous replacement II. J Arthroplasty 11(4):419–428
52. Jacobsen S, Jensen FK, Poulsen K, Sturup J, Retpen JB (2003) Good performance of a titanium femoral component in cementless hip arthroplasty in younger patients: 97 arthroplasties followed for 5–11 years. Acta Orthop Scand 74(4):375–379
53. Yoshinari M, Matsuzaka K, Inoue T, Oda Y, Shimono M (2003) Effects of multigrooved surfaces on fibroblast behavior. J Biomed Mater Res A 65(3):359–368

54. Matsuzaka K, Yoshinari M, Shimono M, Inoue T (2004) Effects of multigrooved surfaces on osteoblast-like cells in vitro: scanning electron microscopic observation and mRNA expression of osteopontin and osteocalcin. J Biomed Mater Res A 68A(2):227–234

55. Matsuzaka K, Walboomers XF, Yoshinari M, Inoue T, Jansen JA (2003) The attachment and growth behavior of osteoblast-like cells on microtextured surfaces. Biomaterials 24(16):2711–2719

56. Chang CK, Wu JS, Mao DL, Ding CX (2001) Mechanical and histological evaluations of hydroxyapatite-coated and noncoated Ti6Al4 V implants in tibia bone. J Biomed Mater Res 56(1):17–23

57. Eckardt A, Aberman HM, Cantwell HD, Heine J (2003) Biological fixation of hydroxyapatite-coated versus grit-blasted titanium hip stems: a canine study. Arch Orthop Trauma Surg 123(1):28–35

58. Mouzin O, Soballe K, Bechtold JE (2001) Loading improves anchorage of hydroxyapatite implants more than titanium implants. J Biomed Mater Res 58(1):61–68

59. Ricci JL, Spivak JM, Blumenthal NC, Alexander H (1991) The bone-biomaterial interface. Modulation of bone ingrowth by surface chemistry and roughness. University of Toronto Press, Ontario

60. Niki M, Ito G, Matsuda T, Ogino M (1991) The bone-biomaterial interface. Comparative push-out data of bioactive and non-bioactive materials of similar rugosity. University of Toronto Press, Ontario

61. Uchida A, Nade SM, McCartney ER, Ching W (1984) The use of ceramics for bone replacement. A comparative study of three different porous ceramics. J Bone Joint Surg Br 66(2):269–275

62. Yamamuro T, Takagi H (1991) The bone-biomaterial interface bone bonding behaviour of biomaterials with different surface characteristics under load-bearing conditions. University of Toronto Press, Ontario

63. Jinno T, Kirk SK, Morita S, Goldberg VM (2004) Effects of calcium ion implantation on osseointegration of surface-blasted titanium alloy femoral implants in a canine total hip arthroplasty model. J Arthroplasty 19(1):102–109

64. Gong W, Abdelouas A, Lutze W (2001) Porous bioactive glass and glass–ceramics made by reaction sintering under pressure. J Biomed Mater Res 54(3):320–327

65. Kato H, Neo M, Tamura J, Nakamura T (2001) Bone bonding in bioactive glass ceramics combined with a new synthesized agent TAK-778. J Biomed Mater Res 57(2):291–299

66. Kato H, Nishiguchi S, Furukawa T, Neo M, Kawanabe K, Saito K, Nakamura T (2001) Bone bonding in sintered hydroxyapatite combined with a new synthesized agent, TAK-778. J Biomed Mater Res 54(4):619–629

67. Capello WN, D'Antonio JA, Manley MT, Feinberg JR (1998) Hydroxyapatite in total hip arthroplasty. Clinical results and critical issues. Clin Orthop Relat Res 355:200–211

68. RÃ¸kkum M, Reigstad A (1999) Total hip replacement with an entirely hydroxyapatite-coated prosthesis: 5 years' follow-up of 94 consecutive hips. J Arthroplasty 14(6):689–700

69. Ciccotti MG, Rothman RH, Hozack WJ, Moriarty L (1994) Clinical and roentgenographic evaluation of hydroxyapatite-augmented and nonaugmented porous total hip arthroplasty. J Arthroplasty 9(6):631–639

70. Rothman RH, Hozack WJ, Ranawat A, Moriarty L (1996) Hydroxyapatite-coated femoral stems. A matched-pair analysis of coated and uncoated implants. J Bone Joint Surg Am 78(3):319–324

71. Johnston DW, Davies DM, Beaupre LA, Lavoie G (2001) Standard anatomical medullary locking (AML) versus tricalcium phosphate-coated AML femoral prostheses. Can J Surg 44(6):421–427

72. Kim YH, Kim JS, Oh SH, Kim JM (2003) Comparison of porous-coated titanium femoral stems with and without hydroxyapatite coating. J Bone Joint Surg Am 85-A (9):1682–1688

73. Yee AJ, Kreder HK, Bookman I, Davey JR (1999) A randomized trial of hydroxyapatite coated prostheses in total hip arthroplasty. Clin Orthop Relat Res 366:120–132

74. Kang JS, Dorr LD, Wan Z (2000) The effect of diaphyseal biologic fixation on clinical results and fixation of the APR-II stem. J Arthroplasty 15(6):730–735

75. Hayashi K, Mashima T, Uenoyama K (1999) The effect of hydroxyapatite coating on bony ingrowth into grooved titanium implants. Biomaterials 20(2):111–119
76. Szmukler-Moncler S, Salama H, Reingewirtz Y, Dubruille JH (1998) Timing of loading and effect of micromotion on bone-dental implant interface: review of experimental literature. J Biomed Mater Res 43(2):192–203
77. Maloney WJ, Jasty M, Burke DW, O'Connor DO, Zalenski EB, Bragdon C, Harris WH (1989) Biomechanical and histologic investigation of cemented total hip arthroplasties. A study of autopsy-retrieved femurs after in vivo cycling. Clin Orthop Relat Res 249:129–140
78. Engh CA, O'Connor D, Jasty M, McGovern TF, Bobyn JD, Harris WH (1992) Quantification of implant micromotion, strain shielding, and bone resorption with porous-coated anatomic medullary locking femoral prostheses. Clin Orthop Relat Res 285:13–29
79. Bernakiewicz M, Viceconti M, Toni A (2001) Investigation of the influence of periprosthetic fibrous tissue on the primary stability of uncemented hip prosthesis. Comput Meth Biomech Biomed Eng 3:21–26
80. Viceconti M, Monti L, Muccini R, Bernakiewicz M, Toni A (2001) Even a thin layer of soft tissue may compromise the primary stability of cementless hip stems. Clin Biomech (Bristol, Avon) 16(9):765–775
81. Weinans H, Huiskes R, Grootenboer HJ (1993) Quantitative analysis of bone reactions to relative motions at implant-bone interfaces. J Biomech 26(11):1271–1281
82. O'Toole Iii RV, Jaramaz B, DiGioia Iii AM, Visnic CD, Reid RH (1995) Biomechanics for preoperative planning and surgical simulations in orthopaedics. Comput Biol Med 25(2):183–191
83. Huiskes R, Verdonschot N, Nivbrant B (1998) Migration, stem shape, and surface finish in cemented total hip arthroplasty. Clin Orthop Relat Res 355:103–112
84. McNamara BP, Cristofolini L, Toni A, Taylor D (1997) Relationship between bone-prosthesis bonding and load transfer in total hip reconstruction. J Biomech 30(6):621–630
85. Stolk J, Maher SA, Verdonschot N, Prendergast PJ, Huiskes R (2003) Can finite element models detect clinically inferior cemented hip implants? Clin Orthop Relat Res 409:138–150
86. Tanner KE, Yettram AL, Loeffler M, Goodier WD, Freeman MAR, Bonfield W (1995) Is stem length important in uncemented endoprostheses? Med Eng Phys 17(4):291–296
87. Prendergast PJ (1997) Finite element models in tissue mechanics and orthopaedic implant design. Clin Biomech 12(6):343–366

Chapter 2
Finite Element Model Construction

Abstract This chapter explains the methodology to simulate total hip arthroplasty. Three-dimensional (3D) model of a femur was created from the Visible Human Project computed tomography dataset. Four different models of cementless hip stem were constructed from various file formats. Triangular surface mesh was manually repaired to ensure good finite element model for analyses. The implant surface mesh was then aligned in the femoral canal and the complete arthroplasty model was then converted into solid tetrahedrals. The implant was assigned with linear isotropic properties and the bone was assigned based on their greyscale values. Loads simulating the gait cycle and stair-climbing were used for the simulation. An algorithm to calculate implant-bone relative motion was developed to analyse the interface micromotion. A convergence study was performed and the micromotion algorithm verified.

Keywords Three-dimensional modelling • Finite element mesh • Hip arthroplasty simulation • Micro-motion algorithm • Convergence and verification

2.1 Pre-processing

2.1.1 Three-Dimensional Model Construction of the Femur

The construction of 3D models of the hip was done using the AMIRA software (TGS software) which allows semi-automated segmentation of medical images. All images used in this study were obtained from two-dimensional (2D) computed tomography (CT) datasets. These images were stacked in order, with a certain thickness value between them that corresponded to the distances between the CT slices. Segmentation was then carried out manually on each slice by marking the required part on the image (Fig. 2.1). The segmented images were then compiled automatically using the software's marching cubes algorithm, generating a

M. R. Abdul Kadir, *Computational Biomechanics of the Hip Joint*, SpringerBriefs in Computational Mechanics, DOI: 10.1007/978-3-642-38777-7_2,
© The Author(s) 2014

Fig. 2.1 Two-dimensional
slice from a VHP dataset,
showing segmentation of the
right femur

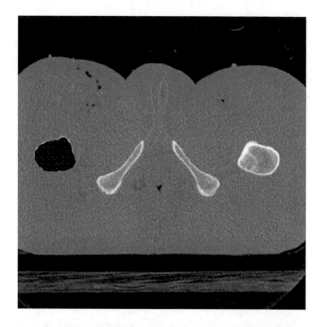

Fig. 2.2 Three-dimensional
model construction with
triangular surface mesh
of the *right femur* from
CT images using AMIRA
software

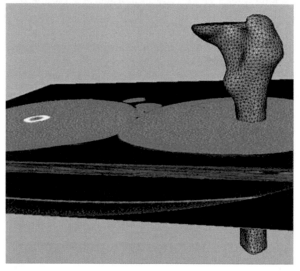

3D triangular surface mesh. The resulting mesh was very refined, and a triangular reduction and a smoothing procedure were performed on the newly created model to turn it into a more manageable mesh density with undistorted triangular shapes (Fig. 2.2). The femoral neck was left uncut for the time being as the anteversion angle is required for orientation of the stem. For most of the project, CT images from the Visible Human Project (VHP) dataset were used as the standard, unless stated otherwise.

The above procedure described the construction of the femur model using 2D image based data in AMIRA. Theoretically, one can also create an implant model from say a CT scan of an implanted femur using the above procedure. However, artefacts from the metal of the implant make it impractical to create the implant model using the above method. This problem will be discussed further in Chap. 4 when analysing the effect of interfacial gaps on micromotion.

2.1.2 Three-Dimensional Model Construction of the Hip Stems

Three-dimensional CAD models of the implants were obtained from their respective manufacturers in various file formats. Creating a mesh from these models depended very much on the format of the file and the available software to do the meshing. Most of the studies here used the AML hip stem (DePuy Orthopaedics, Indiana, US). The files obtained were in a format that was readable in MARC.Mentat (MSC software), the finite element (FE) software used throughout this project. Four sizes of the AML stem were all received in rectangular surface mesh format (Fig. 2.3). The mesh was then refined by subdividing the elements and then converted into a triangular surface mesh by subdividing the rectangular elements using the two corners of each rectangle. Several adjustments were made to the original AML model such as the removal of the collar and tapering

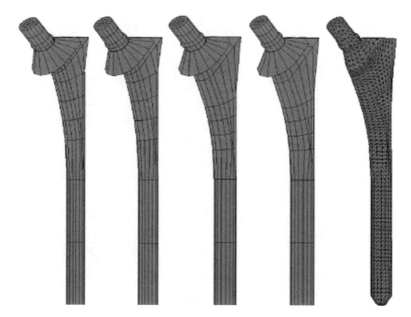

Fig. 2.3 CAD models of the AML stems at different sizes (first 4 stems) received from the manufacturer, and the solid *tetrahedral mesh* created (*right*)

the stem tip. The collar was removed because a study by Keaveny and Bartel [1] found that the collar was not effective in terms of achieving stability and reducing micromotion. The cut bone surface usually resorbs away from contact with the collar. Further discussion on the effects of a collar on micromotion will be made in Chap. 3 when discussing various design aspects of cementless hip stems. The tip was tapered because the actual design has a tapered tip and, more importantly, to eliminate unnecessary stress concentration in this area. The triangular mesh was then automatically turned into solid tetrahedrals using the tetrahedral mesh generator in AMIRA.

The method described above for the AML hip stem was rather straightforward. This is because the models supplied by DePuy were already properly meshed. However, for other 3D models of hip stems used in a later study (Chaps. 3 and 4), they were not in the same format as the AML models. Figure 2.5 below shows the original 3D CAD model of the Alloclassic hip stem (Sulzer Orthopaedics, Switzerland) received in a stereolithographic (.stl) format, a format that is generally used as a pre-processing format for rapid-prototyping technologies [2]. The surfaces were remeshed because the elements were not of equal size and were highly distorted. Another computer software called MAGICS (Materialise Software) was used to remesh the surface, and the process was carried out manually "section-by-section", i.e. removing a plane section and remeshing it with a constant triangular grid of a certain size (Fig. 2.4). This technique was

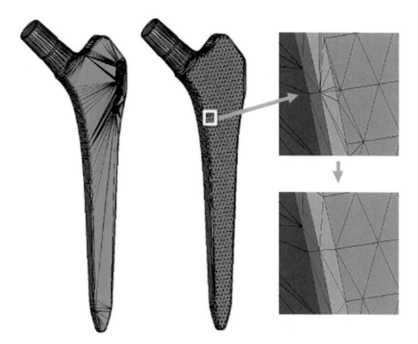

Fig. 2.4 The original CAD drawing with highlighted section for re-meshing (*left*), and after re-meshing (*centre*). The inset pictures show the extra elements created (*top*) and after manual repair (*bottom*)

Fig. 2.5 The original CAD drawing of an Alloclassic Hip Stem (*left*) received from the manufacturer and the completed triangular surface mesh

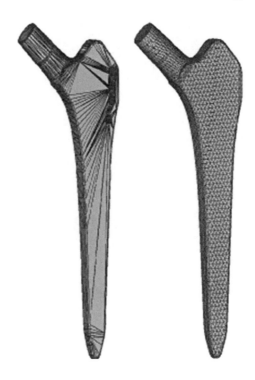

implemented to ensure relatively constant triangular mesh size throughout while maintaining the actual geometry of the implant.

The major problem of using this procedure, however, was that it created extra elements on all intersecting nodes at the edge of the plane, resulting in distorted thin elements at the boundaries of the sections. These unwanted elements were manually checked and repaired on each converted section. Another limitation of this procedure was that it only worked on flat surfaces, so cylindrical geometry such as the neck of the implant needed to be remeshed using a different method; the elements were subdivided in the FE software MARC.Mentat to achieve constant and undistorted elements, and then repaired manually in MAGICS. The above procedure was important in order to maintain the actual geometry of the implant as close as possible and at the same time minimising local stress concentration during FE analysis. The resulting mesh is shown in Fig. 2.5.

The other type of hip stem used in a later study (Chap. 3) was the ABG (Stryker Howmedica Osteonics, Rutherford, NJ), and the 3D solid models of this stem were obtained in IGES format—a geometric data file format than can be imported by CAD software. Creating a surface mesh from this type of file format could be done using the automatic meshing routine in MARC but this option was uncontrollable; it created an uneven mesh density. The only way to create a relatively constant mesh size was to create a triangular mesh "section-by-section", similar to the one used for the.stl file format. The difference is that in IGES format, the overall geometry of the stem was created based on surfaces (Fig. 2.6). These

Fig. 2.6 The ABG hip stem
in the original IGES format
surface meshed in MARC
(*left*) and the completed
model after repairing the
surface boundaries in
MAGICS (*right*)

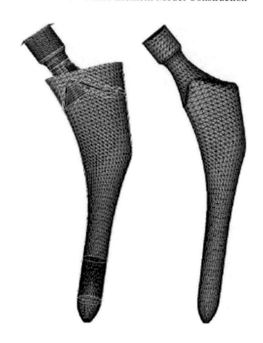

surfaces were meshed individually in MARC with a certain triangular grid size
and MAGICS was used to repair manually the elements across the surface boundaries. This was a highly time consuming process, but produced a properly-spaced
and equally-sized surface mesh.

All triangular surface meshes of the implants were then turned into solid tetrahedral automatically in AMIRA software. The procedure described above was
not the only method for creating 3D FE models and was very time-consuming, but
based on the available software at hand, the method described above was the only
way to create good FE solid meshes.

2.1.3 Alignment of the Hip Stems

Once the construction of the 3D models of the femur and the implants was completed, the next step was to position the implant inside the bone to simulate hip
replacement. The surface mesh, not the solid mesh, of both the implant and the
bone were used for creating bone-implant contact. The simulation of stem insertion into the femoral canal was performed according to the recommended surgical procedures of the particular implant. There are similarities in the principles of
positioning these hip stems. The distal half of the stem and the canal axis must
be aligned, and the neck of the prosthesis must align to the anteversion angle of
the neck of the femur. For the AML prosthesis, the stem and the canal axis were
aligned, with the stem axis being defined as the centre line of the cross section in

the distal portion of the stem, and the canal axis was defined as the centre line of the isthmus region. The isthmus region was assumed to be almost cylindrical in shape, and the maximum inscribed circle on the endosteal geometry was used as a reference for a suitable size of the stem. This is done because the AML is designed for canal filling, where canal fill is defined in cross sectional slices as a proportion between the stem area and the endosteal area of the isthmus. The stem was then rotated so that the neck of the stem was aligned to the anteversion angle of the femoral neck. The osteotomy level was then set at 10 mm above the upper end of the lesser trochanter and coinciding with the osteotomy line of the stem—for the AML it had to be in line with the placement of the collar on the medial calcar.

For the Alloclassic hip stem, similar procedures were used to align the stem and the neck. However, the Alloclassic is rectangularly tapered and was not designed to fill the canal in the medio-lateral (ML) direction. A suitable size was chosen based on the size that appropriately filled the canal in the ML direction with the lateral flare of the stem used as a reference to ensure proximal fill. Contact to the endosteal cortex in the distal half was also used to determine suitable size. As the stem was tapered, maximum cortical support like the one obtained for the AML was not possible.

Similar procedures were used for the ABG hip stem for alignment with the canal and the femoral neck. The ABG is an anatomical stem with distal endosteal bone over-reaming as a standard surgical protocol. The suitable size was therefore decided based on a proximal fill of the stem in the greater trochanter area.

Another type of cementless stem used in this analysis was the CLS hip stem. It is rectangularly tapered in all planes with a significant reduction in diameter in the distal half. Apart from using similar procedures to align the stem and the neck, the lateral part of the stem must also touch the endosteal lateral cortex. The CLS is also a proximal fixation design, and as such a suitable size was decided based on the stem that appropriately filled the proximal part of the femur.

The aligned and appropriately sized stems were presented to and judged by an experienced orthopaedic surgeon to confirm that the reconstruction followed the design concept and implantation of the individual hip stem. Pictures of the aligned and appropriately sized cementless hip stems used in this study are shown in Fig. 2.7.

2.1.4 Construction of Bone-Implant Contact

Once the stem was appropriately in place, the osteotomy level was set to about 10 mm above the upper end of the lesser trochanter. The cut was done using a 3D cube model oriented to the same anteversion angle of the femur using MAGICS (Fig. 2.8). The boolean operation, however, created extra unwanted elements at the boundaries between the bone and the cube model. These bad elements were repaired manually as described before.

To create contact between the bone and the implant, an assumption was made that there would be a perfect fit between the outer surface of the implant and the

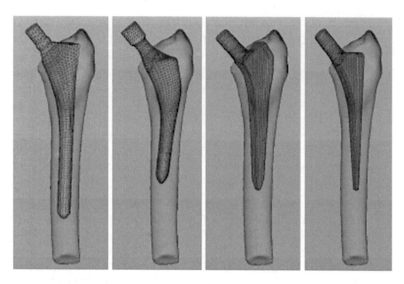

Fig. 2.7 Hip stems used in the analysis—the AML, the ABG, the Alloclassic and the CLS—aligned inside the VHP femur model

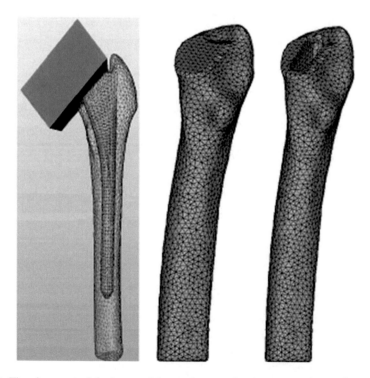

Fig. 2.8 The placement of the box model to make a cut for the ante-version angle at the neck (*left*), the femur model with ante-version cut (*middle*) and the "after-reamed" femur model (*right*)

inner bone. Creating a perfect match at the interface was done by using the surface mesh of the aligned stem as a "cut-out" for the resected hip model using the same boolean operation in MAGICS. The result of this operation was a resected hip model with a hole that had the exact shape of the outer surface of the implant. As with the cut of the neck, there were bad elements created at the boundaries between the implant and the bone, and these elements were repaired manually. The bone surface mesh proximal to the lateral shoulder of the stem was also removed to simulate the actual results of reaming by the surgeon. The "operated" hip model was then automatically turned into solid tetrahedrals in AMIRA.

2.2 Contact Modelling

There are currently two methods of modelling contact; with or without interface elements. Interface elements, sometimes called gap elements, are specialised elements that bridge the displacement field discontinuity caused by frictional sliding and tensile separation. These elements allow force transfer between pairs of interface nodes. The second method, the "direct contact", on the other hand, does not employ structural elements to address the displacement field discontinuity. In this alternative method, a constraint is automatically imposed when a node contacts another node or surface. The constraint equation is such that the contacting node is forced to be on the contacted surface and allowed to slide, subject to the current friction conditions and the calculated tolerance zone. If a node slides from one segment to another during the iteration procedure, the retained nodes associated with the constraint are changed and a recalculation is automatically made. For tensile separation, a default value of 10 % of the maximum reaction force is usually used as the threshold limit.

Both techniques of modelling contact have been used in FE micromotion studies with results comparable to their experiment equivalent. However, there have been reports of potential limitation of using the interface elements. Zachariah and Sanders [3] reported that in contact analyses between soft tissues and stiff surfaces, the direct contact method was sensitive to the coefficient of friction and better reflected the effects of local shape differences. In another study comparing gap elements and zero interface thickness elements, it was found that micromotion patterns obtained using both techniques were similar, but with higher magnitudes in the case of gap elements [4]. In a study comparing the use of gap elements and the direct contact procedure, Viceconti et al. [5] showed that gap elements produced more errors than the alternative method. Furthermore, it has been reported [6] that when friction was simulated in a contact analysis, the predicted relative displacements were dependent on the value of the assigned axial stiffness of the gap.

Due to some potential limitations of gap elements described above, a direct frictional contact method will be used in this study to analyse micromotion. The procedure described previously for reconstruction of 3D models created a perfect fit between the stem and the bone, i.e. the interface nodes of the stem and their corresponding nodes at the bone shared the same co-ordinates. This is appropriate for the direct contact method, and the FE software used in this analysis, MARC, claimed that the direct contact procedure can be very accurate if perfect contact is already known. A description of frictional contact and the study of the effects of friction coefficient on micromotion will be analysed in Chap. 3.

2.3 Material Properties Assignment

Once the solid tetrahedral mesh of the operated femur is completed, material properties will have to be assigned for the implant and the bone. This is rather straightforward for the implant as it is a homogeneous material. Bone, however, is not homogeneous. The mechanical properties of bones vary depending on location within the bone, age, level of activity and pathological conditions. Early FE models defined the properties by separating the stiff cortex from the cancellous structure and assigned a homogeneous property to each of them [5, 7–10]. Cancellous stiffness values used ranged from 0.07 to 0.75 GPa, whilst cortical stiffness was from 14 to 17 GPa. Others used a more sophisticated approach by assigning material properties on an element-by-element basis [1]. The properties were estimated from photographs or radiographs by estimating grey level values.

In this numerical simulation, the second method of assigning material properties was chosen. An in-house algorithm written by previous author [11] was used to assign material properties based on medical imaging data. In this algorithm, the grey-level of the CT images was related to the apparent density using a linear correlation [12, 13]. This allowed for the transformation of the spatial radiological description into the description of bone density. The modulus of elasticity of individual elements was then calculated from the assigned apparent densities using the cubic relationship proposed by Carter and Hayes [14]:

$$E = c\rho^3$$

where $c = 3{,}790$ MPag^{-3} cm^9. This relationship was based on the assumption that cancellous and cortical bone.s are simply at different ends of a continuous spectrum. The material properties were assumed to be linear elastic and isotropic with Poisson's ratio set to 0.35.

2.4 Boundary Conditions

Once the model has been completed, loads simulating physiological activities need to be assigned before analyses could be performed. Loading configurations were different from author to author depending on several factors such as the activities analysed, the pathological condition of the patient and the weight of the patient. This analysis will use two loading configurations, Fisher's gait cycle and Duda's

Fig. 2.9 Coordinate system
used in all analyses (taken
from "Hip 98" CD)

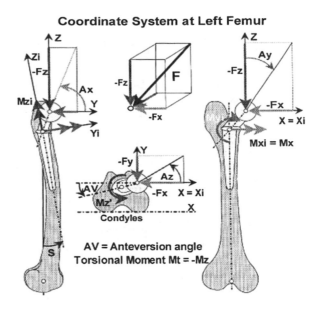

stair-climbing. A detailed description of the two physiological activities and their
effect on micromotion will be discussed in Chap. 4, together with a study on the
effects of muscle forces.

Before these loads could be assigned, a coordinate system needs to be defined.
It is important to use the same sign convention as the coordinate system used in
other studies. Both Fisher and Duda used the coordinate system shown in Fig. 2.9.
The FE model of an implanted femur was then aligned to the coordinate system
before loading vectors of the joint contact force and muscle forces were placed
through point loads. In all cases, the model was constrained at the distal part of the
bone in all directions to prevent rigid body motion.

Several parameters needed to be set such as the coefficient of friction and the
interference fit. A value of 0.4 for the friction coefficient and a radial interference fit of 0.1 mm were usually used in most of the analyses. The reasoning for
these choices and the effects of these parameters on interface micromotion will
be looked at in detail in Chap. 4. Numerical parameters such as the incremental
load step will be discussed later on in this chapter. An FE software program called
MARC.Mentat (MSC Corporation, USA) was used for all micromotion analyses
in this numerical simulation.

2.5 Implant Stability Subroutine

This section describes how micromotion and the stability of the implant were calculated and displayed. One major difference to note between FE predictions and experimental study on relative micromotion is exactly where the measurement of relative

motion is made. Experimental micromotion usually involves displacement transducers mounted on the outer cortex of the bone, i.e. they are separated from the implant by the thickness of the bone. The deflection between the implant and the bone during loading must involve movement at the interface and could also involve deformation of the bone between the surface of the implant and the outer surface of the bone where the transducer is mounted [15]. An FE micromotion study, on the other hand, calculates how much sliding occurs at the interface by subtracting the nodal displacements at the outer surface of the stem from the nodal displacements of the bone surface in contact with the stem. The difference in terms of micromotion results between the experimental and FE methods is analysed later in the chapter (see Sect. 2.7).

A computer code was written in Compaq Visual Fortran (Compaq Computer Corporation) to calculate and display micromotion from MARC.Mentat's post-processing file. Another subroutine was written to automatically store the interfacial nodes—the nodes on the implant surface and their corresponding nodes on the bone—which were involved in the micromotion calculations. These "common" nodes shared the same coordinates and were used in the main subroutine by subtracting the nodal displacement of the stem from the corresponding nodal displacement of the bone. The results obtained from the calculation were then displayed as contour plots on the outer surface of the prosthesis.

As mentioned at the start of the chapter, two sets of results are presented in this study. The first one is simply the contour plots of micromotion obtained from the computer code. The results are usually compared with each other, and those with lesser micromotion are assumed to be better than those with larger micromotion. This technique has been used previously [1, 10], although most published FE work on micromotion has simply presented average values of micromotion in various sections of the stem (e.g. proximal, middle and distal) [7–9, 16].

The above micromotion results are suitable for the comparison of micromotion between implants. However, conclusive evidence could not be made from these results about the stability of different hip stems. An orthopaedic surgeon, for example, would like to know if stability is compromised for a particular hip stem design subjected to a particular physiological loading. This question could not be answered simply by presenting the comparative micromotion results. A novel technique is therefore proposed to predict hip stem instability, where bone loss is simulated at the interface where micromotion exceeds the threshold limit for bone ingrowth.

As mentioned in the previous chapter, the threshold value of relative micromotion, above which fibrous tissue layer forms, varies between 30 and 150 μm. In this analysis, a threshold value of 50 μm was chosen. The result from the first implant loading iteration was modified by creating a gap at the interface where micromotion was found to be more than 50 μm (Fig. 2.10). The 500 μm thickness gap was created by translating the nodes on the bone side of the interface perpendicularly outwards from the stem's surface. A second iteration was then performed with the newly-created interfacial gaps. From the new set of results, the model was modified again to include new gaps at the interface where micromotion exceeded the threshold limit. Third iterations were performed and the procedure was repeated until either a stable-state interface micromotion was achieved or the implant failed.

Fig. 2.10 Pictures showing no interface gaps (*left*) and with interfacial gaps with thickness of 500 μm (*right*)

Implant failure occurred if interfacial shear strength was exceeded or the surface of the implant was encapsulated with the threshold micromotion limit representing fibrous tissue. Interfacial shear strength varied from 6 to 55 MPa based on implant push-out tests [17]. This variation depended on many factors such as implant material, surface texture, bone type and follow-up period. In this numerical simulation, a value of 15 MPa was assumed as the interfacial shear strength based on the published data for a titanium alloy implant with cortical bone type.

In this analysis, either the 'as implanted' or both 'as implanted' and 'post interface remodelling' results will be presented depending on the situation. When looking at the effects of parameters such as the effect of muscle forces on micromotion (such as the one in Chap. 3), then only comparative micromotion results are presented. In Chap. 4, however, when looking at the effect of hip stem designs on micromotion, both sets of results will be presented.

2.6 Convergence Study

The results of any FE analyses must be independent of all purely numerical parameters. It is therefore crucial to perform convergence studies to ensure that the results obtained are not dependent on those parameters. In this section, two parameters will be checked for convergence—the mesh and the load increment.

2.6.1 Mesh Convergence

There are many element types available in finite element analysis, but the discussion will concentrate on element types to model three-dimensional solids. Two of the most common ones are the hexahedral and the tetrahedral, which can be further, categorised into lower order and higher order depending on the interpolation

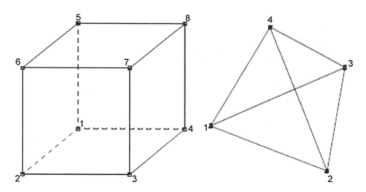

Fig. 2.11 The eight-noded hexahedral (*left*) and the four-noded tetrahedral (*right*). Higher order hexahedrals and tetrahedrals have an extra node (*midside node*) on each edge

function used. A lower order hexahedral is made up of 8 nodes and 8 integration points whilst its tetrahedral companion has 4 nodes and 1 integration point (Fig. 2.11). Both elements have three global displacement degrees of freedom and use linear interpolation functions, where the strains are constant throughout the element. A higher order hexahedral has 20 nodes, 27 integration points and a higher order tetrahedral has 10 nodes and 4 integration points. Both of them can use a quadratic interpolation function which allows the strains to vary within the element.

Higher order elements are usually not recommended in contact analyses because the shape functions have interpolation functions which lead to the equivalent nodal forces oscillating between the corner and midside nodes. As this has a detrimental effect on both contact detection and determining contact separation, lower order elements are usually recommended in contact analyses. The choice of the type of element (hexahedral or tetrahedral) should not, in principle, change the results because both of them have to be in models with converged solutions. However, the total number of elements (or mesh density) of the model may be different between the two different element types. As the tetrahedrals have fewer nodes and therefore fewer degrees of freedom, and because of the difference in shape functions, more of this type of element is needed than the hexahedrals to produce similar converged results.

In this numerical simulation, tetrahedrals are used mainly due to the availability of computer software, AMIRA and MAGICS, which can only work with triangular surfaces. Therefore a convergence study of this type of element will be carried out. To perform the convergence analyses, an AML hip stem model obtained from the manufacturer with a rectangular surface mesh was refined by subdividing the rectangles. The newly subdivided surface mesh was further subdivided until five models with increasing mesh surface density were obtained. These five models of rectangular surface mesh were then converted into triangular surface meshes before converting them into solid tetrahedrals using the automatic solid meshing function in MARC.Mentat as shown in Fig. 2.12. The bone models also followed the same procedure for refinement. Table 2.1 shows the total number of elements

Fig. 2.12 The tetrahedral elements of the implant with increasing mesh densities

Table 2.1 Total number of nodes and elements of tetrahedral mesh with increasing density

	Nodes	Element
Tet-A	4,049	14,688
Tet-B	6,663	28,732
Tet-C	9,225	40,408
Tet-D	12,078	56,526
Tet-E	16,866	77,738

and nodes for all five models. Micromotion analyses were performed on all of the models with all other parameters being the same.

From the contour plot of micromotion shown in Fig. 2.13, sixteen nodes along the length of the stem from the coarsest mesh (Tet-A) were chosen for comparison between all five models. These nodes are available in all models at exactly the same location, thus allowing proper and exact comparison. The graph of micromotion of these sixteen nodes is shown in Fig. 2.14.

There was a significant difference between the coarsest (Tet-A) and the rest of the refined models. However, once refinement was made, there was little difference between all four refined models. The tetrahedral model seemed to converge after two refinements (Tet-C) with a maximum variation of 8 %. This was roughly about 10,000 nodes and 40,500 tetrahedral elements. Ideally, one would do a convergence study for each analysis performed. However, this would not be practical especially in later chapters where various types of implants and bone types were analysed. Therefore, the total number of nodes and elements required for convergence which was concluded in this section will be used as a guideline for convergence in later studies.

Fig. 2.13 Micromotion
results at increasing
tetrahedral mesh densities
(*left* to *right*)

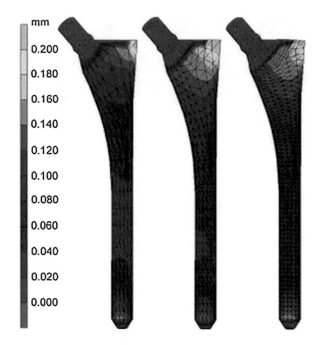

Fig. 2.14 Graph of
micromotion along the axis
of the stem for different mesh
densities

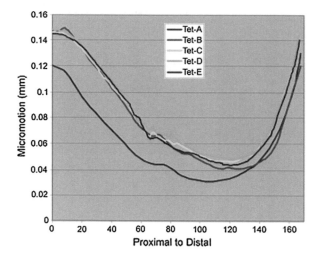

2.6.2 Load Increment

The second FE parameter that could affect the result is the load increment size. The
default setting for load increment in MARC.Mentat is 10, which means that the
model is loaded with an increment of 200 N on each load step if the total load is
2 kN. Three models were prepared from the previous mesh convergence study with

different load increment sizes of 1, 5 and 10 (Fig. 2.15). Micromotion analyses were
then performed with all models having the same loading conditions as well as material properties and friction coefficient. Similar distribution was found when 5 or 10
increments were used as depicted in Fig. 2.16. The difference between these two
increments was 2 % and therefore, 5 load increments are used throughout the study.

Fig. 2.15 Contour plot of
micromotion for load step of
1, 5 and 10 increments

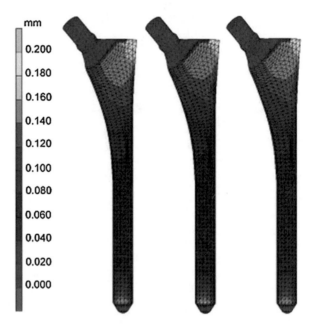

Fig. 2.16 The micromotion
distribution with 1 increment,
5 and 10 increments

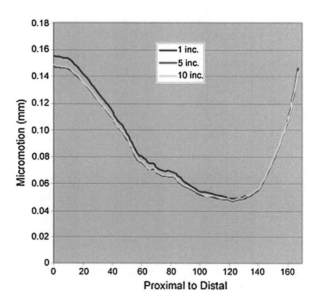

2.7 Verification of Micromotion Algorithm

In this section, the micromotion algorithm was checked for accuracy using a simplified cylindrical bone-implant model. This model was also used for verification purposes, where micromotion results for an implant under a bending moment were compared with the published FE results for a similar model. Finally, the model is used to study the effect of location for measurement of relative micromotion, i.e. comparing the difference between experimental relative micromotion and FE interface micromotion.

2.7.1 Rohlmann's Model

The geometry used for the simplified cylindrical model was taken from the work of Rohlmann et al. [18]. It was a three-dimensional finite element model of a porous coated cylindrical implant inside a femoral diaphyseal bone. A tube with an outer diameter of 30 mm, inner diameter of 20 mm and a total length of 184 mm was used to model the diaphyseal bone. The prosthetic stem was a solid cylinder with a diameter of 17 mm. The thickness of the porous layer was 1.5 mm. The model was loaded by a pure bending moment of 1 Nm at the proximal end of the stem and clamped on the distal end of the bone. The line A–B represents the lateral side and the line C–D the medial side (Fig. 2.17). The assigned elastic moduli were 18 GPa for the bone, 200 GPa for the prosthetic stem and 5 GPa for the porous coating.

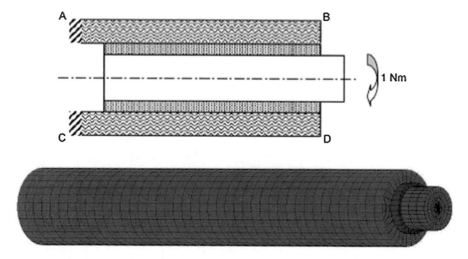

Fig. 2.17 A diagrammatic sketch of a longitudinal section of Rohlmann's model and the finite element model

Rohlmann's FE model consisted of 1,110 eight-noded hexahedral elements and 1,582 nodes. The FE calculations were performed using an FE program called ADINA and a specially developed code to represent the situation directly after implantation, when no tensile stress can be transferred across the bone-implant interface. The code had a nonlinear interface condition where the connection between the elements in the interface was eliminated when tensile stress perpendicular to the interface occurred. A new FE calculation was then performed, thus allowing changes in the stress distributions. This process was repeated several times until only small changes in the stresses were found.

The FE model used in this comparative study consisted of 14,940 eight-noded hexahedral elements and 16,638 nodes. All other parameters (geometry, material properties, boundary conditions) were the same as Rohlmann's. The FE software used was MARC.Mentat, with its default contact algorithm as described in Sect. 2.2, and a friction coefficient of 0.4.

2.7.2   Comparative Analysis

Two sets of results are presented. The first is the accuracy check, where the bending moment and the contact algorithm were removed (i.e. the stem can slide through the bone). A longitudinal displacement of 4 μm and a rotational displacement of 3 μm were applied to the prosthetic cylinder. The analysis was performed with the micromotion code and the contour plot is shown in Fig. 2.18.

The results from numerical simulation showed that the micromotion code accurately calculated and displayed the interface relative micromotion of 5 μm. Figure 2.19 shows contour plots of micromotion, using the micromotion code, when the model was loaded with a pure bending moment of 1 Nm. Rohlmann's published result is displayed on the right. Figure 2.20 shows the graph of relative motion along the cylindrical stem, where the distribution of micromotion along the stem was similar to Rohlmann's. Peaks were found at both ends at the lateral and medial side, with maximum micromotion at the distal-medial point C. The rest of the surface had relatively small micromotion. The maximum micromotion obtained from the FE model was 0.517 μm compared to the analysis by Rohlmann which was 0.389 μm.

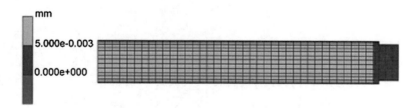

Fig. 2.18 Contour plot showing relative micromotion of 5 μm when the prosthetic stem (with the coating) was displaced longitudinally by 4 μm and rotationally by 3 μm

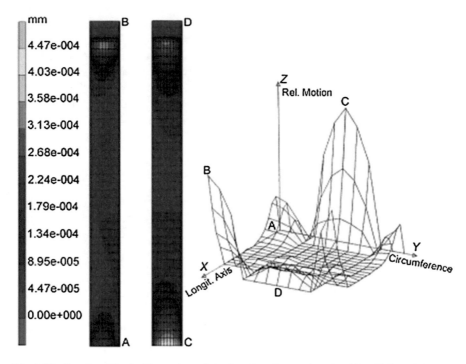

Fig. 2.19 Results of the Rohlmann's model using the micromotion algorithm (*left*) and the published result by Rohlmann et al. (*right*)

Fig. 2.20 Graph of relative motion along the lateral and medial side of the cylindrical prosthesis

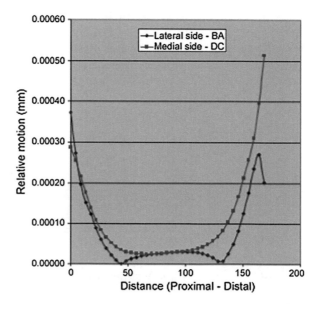

The aim of this particular study was to verify that the micromotion code used in this analysis would give similar results to those published by Rohlmann et al. [18]. The geometrically simplified FE model of Rohlmann was re-created and loaded with a pure bending moment of 1 Nm. The FE results showed excellent agreement with Rohlmann's in terms of micromotion distribution. In the paper, Rohlmann et al. presented only the graph (Fig. 2.19—*right*) and recorded that the maximum micromotion value was 0.389 μm. This value is smaller than our FE result of 0.517 μm. This discrepancy could be attributed to the difference in the FE software used and the specially developed code for the nonlinear interface conditions used by Rohlmann et al. [18].

2.7.3 The Effect of Location for Micromotion Measurement

In this sub-section, the same model was used to study the effect of different locations for micromotion measurement. As discussed previously in Sect. 2.5, one major difference between in vitro experimental micromotion work and FE micromotion prediction is where the motion of the implant was calculated relative to. In experimental studies, displacement transducers are mounted on the outer cortex of the bone, i.e. they are separated from the implant by the thickness of the bone. When the implant is loaded, the deflection between the implant and the transducer involves movement at the interface and could also involve deformation of the bone.

Finite element micromotion studies, on the other hand, are different because relative micromotions are calculated at the interface by subtracting the nodal displacements at the outer surface of the stem from the nodal displacements of the bone surface in contact with the stem. The results could be different from experimental studies if the deformation of the bone is not taken into account.

To the author's knowledge, there are no reports of how micromotion of the implant relative to the bone would seem to change if the measurement was made relative to the periosteal surface instead of the interface. The aim of this analysis is therefore to compare, using a simplified cylindrical FE model, micromotion results at the interface with micromotion of the implant calculated relative to the periosteal surface.

As explained in Sect. 2.5, a short computer code was written to store interfacial nodes—the nodes on the implant surface and their corresponding nodes on the bone—which share the same co-ordinates. This computer code was adjusted so that the nodes on the implant's surface were now paired with the nodes on the outer cortex of the bone directly normal to them. The micromotion code would therefore calculate the difference in the nodal displacements of the implant's surface relative to the outer surface of the cortex.

The bending moment of 1 Nm used previously was increased to 20 Nm to roughly simulate the actual bending moment during physiological loading, with the distal part of the stem fixed in all directions to prevent rigid body motion. The results are shown in Figs. 2.21 and 2.22.

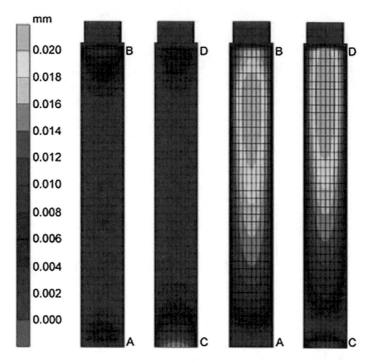

Fig. 2.21 Contour plot of micromotion for the bending moment load of 20 Nm. (*Left* interface micromotion, *right* micromotion relative to the periosteal surface)

Fig. 2.22 Graph of micromotion along the stem on the lateral and medial side under bending moment for the interface micromotion and micromotion relative to the periosteal surface

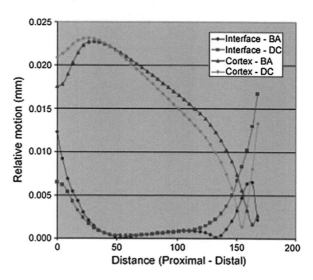

The micromotion calculated relative to the outer surface of the cortex under a bending moment was different from the interface micromotion in terms of magnitude and distribution of micromotion. In areas where interface micromotion was minimum, cortex measurement showed relatively larger micromotion.

The aim of this particular study was to quantitatively analyse micromotion measurement made relative to the cortex with the interface micromotion. Several authors mentioned that bending of the femur could affect the micromotion measurement [15], but to our knowledge there are no reports that quantitatively analysed the hypothesis. Based on a simplified cylindrical model, there were large differences between the interface and cortex measurement when bending deformation of the bone occurred. Since the femur is primarily loaded in bending during physiological loading, care should be taken when comparing between FE and experimental results.

References

1. Keaveny TM, Bartel DL (1993) Effects of porous coating, with and without collar support, on early relative motion for a cementless hip prosthesis. J Biomech 26(12):1355–1368
2. Sanghera B, Naique S, Papaharilaou Y, Amis A (2001) Preliminary study of rapid prototype medical models. Rapid Prototyping J 7(5):275–284
3. Zachariah SG, Sanders JE (2000) Finite element estimates of interface stress in the transtibial prosthesis using gap elements are different from those using automated contact. J Biomech 33(7):895–899
4. Hefzy MS, Singh SP (1997) Comparison between two techniques for modeling interface conditions in a porous coated hip endoprosthesis. Med Eng Phys 19(1):50–62
5. Viceconti M, Muccini R, Bernakiewicz M, Baleani M, Cristofolini L (2000) Large-sliding contact elements accurately predict levels of bone-implant micromotion relevant to osseointegration. J Biomech 33(12):1611–1618
6. Tissakht M, Eskandari H, Ahmed AM (1995) Micromotion analysis of the fixation of total knee tibial component. Comput Struct 56(2–3):365–375
7. Ando M, Imura S, Omori H, Okumura Y, Bo A, Baba H (1999) Nonlinear three-dimensional finite element analysis of newly designed cementless total hip stems. Artif Organs 23(4):339–346
8. Biegler FB, Reuben JD, Harrigan TP, Hou FJ, Akin JE (1995) Effect of porous coating and loading conditions on total hip femoral stem stability. J Arthroplasty 10(6):839–847
9. Kuiper JH, Huiskes R (1996) Friction and stem stiffness affect dynamic interface motion in total hip replacement. J Orthopaed Res 14(1):36–43
10. Viceconti M, Monti L, Muccini R, Bernakiewicz M, Toni A (2001) Even a thin layer of soft tissue may compromise the primary stability of cementless hip stems. Clin Biomech (Bristol, Avon) 16(9):765–775
11. Hopkins AR (2005) Total shoulder arthroplasty simulation using finite element analysis. Dissertation, Imperial College, London
12. Cann CE, Genant HK (1980) Precise measurement of vertebral mineral content using computed tomography. J Comput Assist Tomogr 4(4):493–500
13. McBroom RJ, Hayes WC, Edwards WT, Goldberg RP, White AA 3rd (1985) Prediction of vertebral body compressive fracture using quantitative computed tomography. J Bone Joint Surg Am 67(8):1206–1214
14. Carter DR, Hayes WC (1977) The compressive behavior of bone as a two-phase porous structure. J Bone Joint Surg Am 59(7):954–962
15. Whiteside LA, White SE, Engh CA, Head W (1993) Mechanical evaluation of cadaver retrieval specimens of cementless bone-ingrown total hip arthroplasty femoral components. J Arthroplasty 8(2):147–155
16. Rubin PJ, Rakotomanana RL, Leyvraz PF, Zysset PK, Curnier A, Heegaard JH (1993) Frictional interface micromotions and anisotropic stress distribution in a femoral total hip component. J Biomech 26(6):725–739

17. Dhert WJA, Jansen JA (2000) Mechanical testing of bone and the bone-implant interface. The validity of a single pushout test, 1 edn. CRC Press, USA
18. Rohlmann A, Cheal EJ, Hayes WC, Bergmann G (1988) A nonlinear finite element analysis of interface conditions in porous coated hip endoprostheses. J Biomech 21(7):605–611

Chapter 3
The Effect of Implant Design on Stability

Abstract This chapter analysed the effect of implant design on primary stability. Three different categories were formed from a collection of cementless hip stem designs—straight cylindrical, taper and anatomical. A representative of each category was analysed where similar stress distributions and magnitudes were obtained under the simulated physiological loadings. The three most common materials used for implant—cobalt chromium alloys, titanium alloys and isoelastic composite—were also analysed. The isoelastic stem, though touted as mechano compatible due to its similarity with bone properties, produced a ten-fold increase in relative micromotion. In anticipation of the use of short stem in conservative approach, the finite element analysis showed that a very short stem that covers the proximal region compromised the primary stability. A comparison was also made between the proximal and distal fixation stem where the proximal design was found to be unstable under the gait and stair-climbing activities.

Keywords Cementless hip stems • Straight cylindrical stem • Tapered stem • Anatomical stem • Implant geometry and properties

3.1 The Global Geometry

Cementless hip stems come in different shapes and sizes. In order to analyse practically the effect of these different geometries on primary stability, the stems were grouped into several categories based on their features. There is no consensus at the moment in terms of grouping cementless stems according to their geometry, mostly due to the large variety of cementless stems available in the market today. Healy [1] grouped cementless femoral components into 5 basic types with examples of each— the cylindrical distal filling (the AML, the Solution), the anatomic, proximal fit and fill (the PCA, the Anatomic), combination (the S-ROM, the Bridge), dual, tapered wedge (the Omninfit, the Osteolock, the Mallory-Head, the Synergy, the Summit), and flat, tapered wedge (the Tri-Lock, the Taperloc, the Accolade). Mallory et al. [2]

grouped them into three distinct design geometries and philosophies—the extensive porous coating with distal fixation (the AML), the anatomic proximal fixation (the PCA, the Anatomic) and gradual proximal to distal off-loading tapered geometry (the Mallory-Head). Two other papers reported similar groupings [3, 4].

They grouped the overall geometry into three main categories—cylindrical, tapered and anatomic. Another paper, although it did not categorically group the different types of cementless stems, did mention a comparison between the tapered Taperloc stem to the anatomic PCA and cylindrical stems such as the Harris-Galante, the APR-1 and the AML [5].

A search was conducted in the literature for all follow-up studies and in vitro experimental work to get as much information as possible on the various designs of cementless hip stems. Apart from the literature, nine cementless primary hip stem manufacturers were also included in the search:

1. Aesculap, Tuttlingen, Germany.
2. Biomct, Warsaw, IN.
3. Corin Medical, Gloucestershire, UK.
4. DePuy, Warsaw, IN.
5. Smith and Nephew, Memphis, TN.
6. Stryker Howmedica Osteonics, Rutherford, NJ.
7. S and G Implants, Lübeck, Germany.
8. Wright Medical Technology, Arlington, TN.
9. Zimmer, Warsaw, IN.

From this search, three groups based on the overall geometry of the stem were chosen similar to the one proposed by other authors mentioned above—the tapered design, the anatomic and the straight cylindrical. Hip stems which were not tapered in any plane in the distal half were grouped together. Because all hip stems in this group were also cylindrical, the group was termed straight cylindrical (Fig. 3.1). The tip of the stem may or may not be tapered. Tapered stems were defined as stems that have a proximal to distal taper in either or both the sagittal or longitudinal planes (single-planar/biplanar tapered) (Fig. 3.2). Some tapered designs such as the Mallory-Head also have a posterior-to-anterior taper in the coronal plane (tri-planar tapered).

Anatomic stems were defined as stems designed with an anterior-posterior curve that mimics the natural curve of the human femur (Fig. 3.3). The stems must therefore come in a left and right component. All cementless femoral stems create a press-fit mechanical interface where contact pressures between two components of dissimilar modulus, the bone and the implant, produce deformation. The bone exhibits viscoelastic behaviour, which limits the effectiveness of the press-fit by relaxing the contact pressures at the interface [6]. The cylindrical distal fixation stems rely on cortical support in the distal aspect of the stem for stability, whereas tapered stems rely on proximal, cancellous bone contact and a 3-point fixation pattern. The Anatomic stems employ curved designs in both A-P and M-L

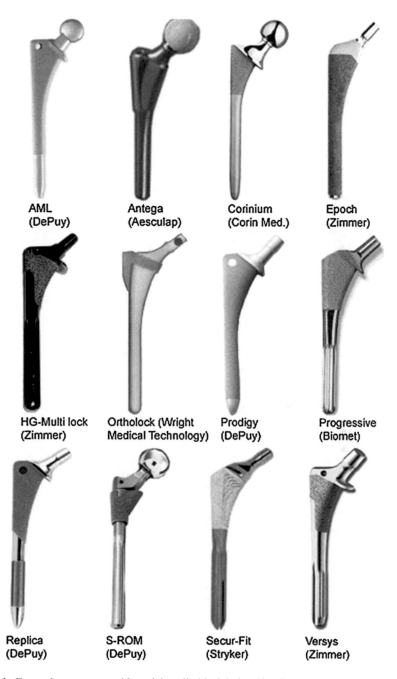

Fig. 3.1 Femoral components with straight cylindrical design (the pictures are not resized to scale)

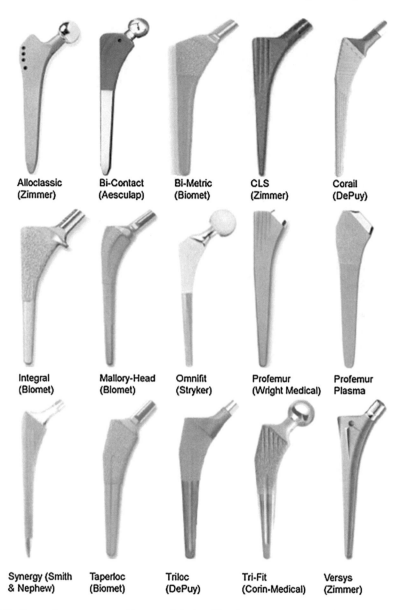

Alloclassic (Zimmer)	Bi-Contact (Aesculap)	Bi-Metric (Biomet)	CLS (Zimmer)	Corail (DePuy)
Integral (Biomet)	Mallory-Head (Biomet)	Omnifit (Stryker)	Profemur (Wright Medical)	Profemur Plasma
Synergy (Smith & Nephew)	Taperloc (Biomet)	Triloc (DePuy)	Tri-Fit (Corin-Medical)	Versys (Zimmer)

Fig. 3.2 Femoral components with tapered design (the pictures are not resized to scale)

planes, and have a large proximal segment to achieve a closer match to the natural endosteal cavity of the proximal part of the femur. It has been claimed that this design feature optimise resistance to axial, bending and rotational forces [7].

The groupings were made based on the overall geometry of the stem. The philosophies may or may not be grouped together. The straight cylindrical design, for example, may have two types of fixation—proximal and distal. The Prodigy

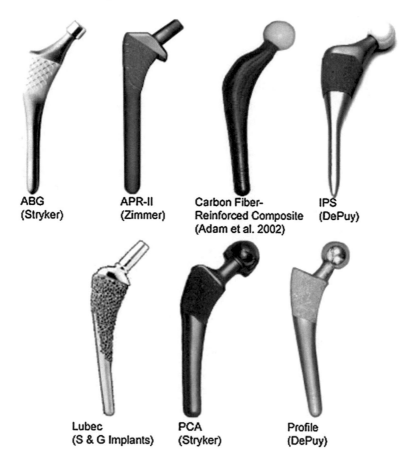

ABG
(Stryker)

APR-II
(Zimmer)

Carbon Fiber-
Reinforced Composite
(Adam et al. 2002)

IPS
(DePuy)

Lubec
(S & G Implants)

PCA
(Stryker)

Profile
(DePuy)

Fig. 3.3 Femoral components with anatomic design (the pictures are not resized to scale)

is a distal fixation design, whereas the AML, though intended for distal fixation, can also work as proximal fixation design [8]. Similarly for the tapered group, the Alloclassic is distally fixed, whereas the CLS is also known to be a proximal fixation design. For the anatomical group, the ABG and the IPS employ a distal bone over-reaming technique for proximal fixation, but the Profile prosthesis, which uses a similar fixation concept, does not use the over-reaming procedure. The issues of proximal and distal fixation concepts will be looked at in Sect. 3.4.

In order to analyse the stability of these hip stems, a member of each group was chosen and analysed. The chosen design was only based on the availability of the 3D CAD model obtained from the respective manufacturer. They are the AML for group 1, the Alloclassic for group 2, and the ABG for group 3.

The Anatomic Medullary Locking (AML) hip stem (Depuy, Warsaw, Indiana) is a nontapered collared design with a relatively longer stem. The shape of the stem is cylindrical to fit the medullary canal and tapered at the tip to help reduce the potential for thigh pain. Because the stem is not tapered, the implant does not wedge in place.

Instead, fixation depends on a so-called 'scratch fit' between the rough external surface of the implant and a similarly shaped bone canal [9]. The stem is porous coated for about 80 % of its length from the proximal end and is polished distally. It is made of cobalt chromium (CoCr) with the porous coating made of sintered CoCr beads with a mean pore size of 200 microns. The pores have been shown to provide a suitable surface to encourage tissue ingrowth and improve the stability of hip stems [10].

The Alloclassic (Zimmer, Warsaw, IN) is a cementless, flat, tapered, collarless hip stem using distal fixation press-fit technique. The surface is fully grit-blasted (4–6 microns) with corundum particles for future bone ongrowth. A better proximal fit is achievable by having a lateral wing engaging the greater trochanter, claimed to improve axial stability. It is made of Protasul-100 titanium alloy (TiA Al Nb) where niobium is used instead of the usual vanadium used in titanium alloy. The straight taper is flat in transverse section and wedged shaped mediolaterally. It is designed to fit the femoral canal in the frontal plane but does not fill in the lateral plane. It achieves self-locking to the endosteal bone by way of 4 corners of its rectangular shape along the entire stem.

The Anatomique Benoist Giraud (ABG) hip stem (Stryker Howmedica Osteonics, Rutherford, NJ) is a cementless anatomic femoral component of roughened titanium alloy with a 50 microns coating of hydroxyapatite applied to its proximal 1/3. A proximal press-fit technique is applied with distal femoral overreaming as standard surgical procedure to avoid bone contact in this area. This design of proximal metaphyseal fixation in the ABG was an attempt to transfer stresses more proximally to maintain bone density.

3.1.1 Finite Element Modelling

Although all three chosen implants—the AML, the Alloclassic and the ABG—have been grouped into different categories of hip stem designs, these implants had other unique characteristics or special features of their own (Fig. 3.4 and Table 3.1). This made the analyses difficult as the specific features of the implants may have interacted with the global geometry and affected the overall primary stability. Therefore, these implants were adjusted, so that a proper comparison based on the groupings above could be made. For the AML, the collar was removed, the stem was shortened to the average length of the Alloclassic and the ABG, and the distinction between the porous-coated and the smooth section was ignored. The prosthesis was regarded as having a homogeneous surface structure throughout the stem. This also applied to the ABG, where the different surface finish between the proximal and distal part was ignored. The indentation features in the proximal part of the ABG were removed. The surgical technique of the ABG requires that the distal part is over-reamed to avoid cortical contact in this area. For the sake of analysing the different groups in this section, bone over-reaming was not modelled. It was assumed that there was perfect contact at the interface between the stems and the bone for all models. The coefficient of friction was set to 0.4 and an interference fit of 0.1 mm was used

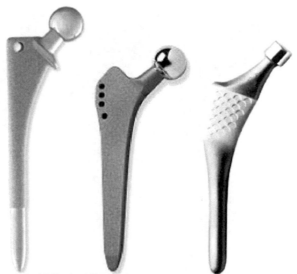

Fig. 3.4 Pictures of the AML, the Alloclassic and the ABG, taken from the manufacturer's website

Table 3.1 The three types of hip stem designs analysed and their characteristics

	Fixation type	Material	Stem length	Stem shape	Symmetry	Surface finish
The AML	Distal	Co Cr	Long	Cylindrical	Yes	Porous-coated
The Alloclassic	Distal	Ti Al Nb	Medium	Rectangular	Yes	Grit-blasted
The ABG	Proximal	Ti Al	Medium	Cylindrical	No	Macrofeature

throughout. The adjusted models will therefore be referred to as the 'cylindrical' (group 1), the 'tapered' (group 2) and the 'anatomical' (group 3) designs respectively. The models were loaded in accordance with both Fisher's gait analysis and Duda's stair-climbing loads and the results were then compared between each other.

From the first set of results, bone elements with a surface area of more than 50 μm of interface micromotion were adjusted so that contact between these elements and the implant was no longer available. This was done to simulate the effect of interfacial bone loss. A detailed description of the method has been explained in Chap. 2. The models were then reloaded with physiological walking and stair-climbing loads to check for instability.

3.1.2 Biomechanical Assessment of Different Hip Stem Designs

Figure 3.5 shows that the magnitudes and distribution of micromotion were similar in all three types of implants, in both physiological walking and stair-climbing. Large micromotions were found in the proximal areas and around the distal stem

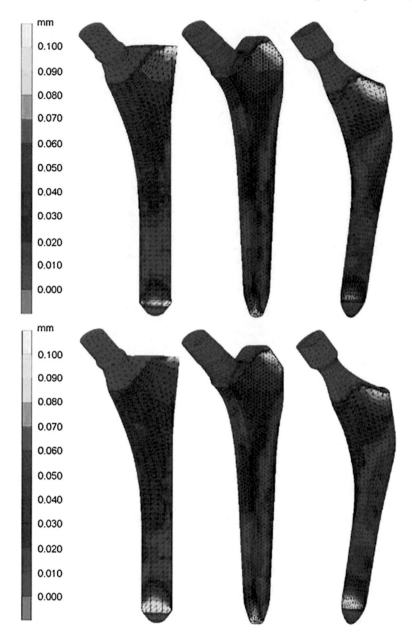

Fig. 3.5 Contour plots of micromotion for the cylindrical (*left*), the tapered (*middle*) and the anatomical (*right*) using Fisher's gait loading (*top*) and Duda's stair-climbing loads (*bottom*) after the 1st iteration

tip. In terms of the amount of surface area with more than 50 μm of interface micro-motion, the range was between 8 and 10 % for all designs. It showed that no specific global design feature was better than the other.

After removing the bones with micromotion in excess of the chosen threshold limit to simulate interfacial bone loss, results were then compared once more (Figs. 3.6, 3.7 and Table 3.2). In general, all designs were found to be stable with bone loss only increased slightly (up to 13 %). The anatomical design was found to be the most stable with a very small increase in surface area above 50 μm. The cylindrical design was the worst in stair-climbing with an increase in unfeasible surface area from 9 to 13 %. The tapered design was the worst in physiological walking where there was an increase from 8 to 10 %.

The three categories of hip stems analysed in this section, the cylindrical, the tapered and the anatomical designs had similar distribution of micromotion and were all stable when bone loss was simulated. The findings are in agreement with published results of actual hip stems belonging to these groups. The AML, the Alloclassic and the ABG are all implants with excellent survival rates in short-term, medium-term and long-term.

The AML is a successful hip stem and performed well in hip arthroplasty [11]. The results are well-published and the studies covered patients with a wide range of ages. One study [12] reported 15 years of clinical experience with the AML. Out of 393 AML stems implanted only 6 have been revised, 3 of which were due to loosening. Another study [13] reported a survival rate of 92 % at 10.5 years for the AML. 88 % of the patients had good or excellent clinical results, but calcar resorption was found in 40 %. One clinical trial included 154 patients under the age of 50, with an average age of 37.6 years [14]. Eighty-eight of the AML implants were followed for at least 10 years, and the authors reported excellent lasting results in these young patients. The reported survival rate was 99, 96 % of which has bone ingrown, 3 % stable fibrous tissue and only 1 % was found unstable.

The FE results presented in this section were also in agreement with other follow-up studies in terms of predicting bone ingrowth. One study [15] reported that when solid initial fixation is obtained intraoperatively and radiographically using the AML stem, bone ingrowth reliably occurs whether or not a partial or full weight-bearing postoperative protocol is followed. Another study [16] reported that bone ingrowth occurred in 93 % of the cases where maximum stability was achieved. An early retrieval study of an AML stem prior to gross failure showed dense cortical and cancellous bone ingrowth [17]. Strength of attachment of the metal implant to bone was good and no slippage was found at the interface when tested under torsional and axial load. Another study used backscattered scanning electron microscopy (SEM) to analyse bone growth of the AML stem [12]. They reported that average bone ingrowth was found on 57 % of the porous-surfaced area of the femoral components.

An experimental study has been conducted comparing a freshly implanted AML stem and a specimen retrieved from a deceased patient [18]. They found that micromotion was greater proximally for the freshly implanted stem compared to the bone-ingrown retrieved specimen. They also found that the smooth distal stem of the retrieved specimen caused larger micromotion than the freshly implanted. The authors concluded that the flexibility of the femur causes increased micromotion of the femur around the smooth distal stem despite the initially tight distal fit.

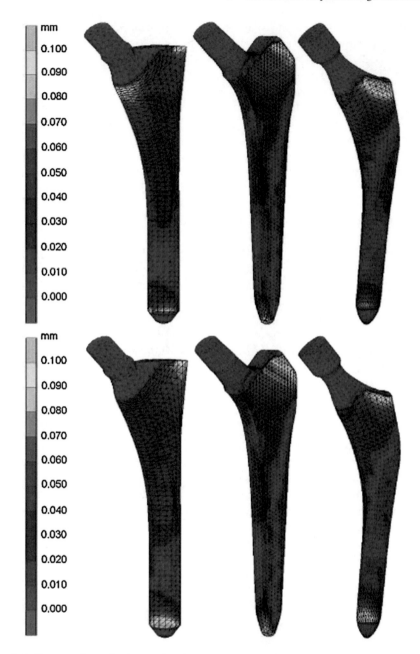

Fig. 3.6 Contour plots of micromotion for the cylindrical (*left*), the tapered (*middle*) and the anatomical (*right*) using Fisher's gait loading (*top*) and Duda's stair-climbing loads (*bottom*) after simulated interfacial bone loss

For the anatomical design, there are also follow-up reports confirming the FE predictions. The ABG hip stem has been reported to have excellent clinical and radiographic results at short-term [19, 20] and in 5 years follow-up [21, 22],

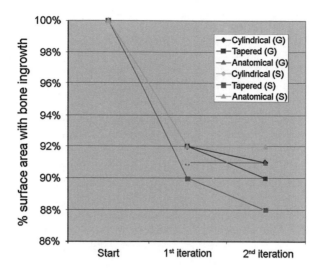

Fig. 3.7 Percentage area of predicted bone ingrowth for the cylindrical, the tapered and the anatomical stem designs using Fisher's gait (G) and Duda's stair-climbing (S) loads

Table 3.2 Surface area > 50 μm of micromotion for the cylindrical (7,345 mm^2), the tapered (7,690 mm^2) and the anatomical (7,222 mm^2) after 1st, 2nd and 3rd iterations

		First iteration		Second iteration		Third iteration	
		Area > 50 μm (mm^2)	(%)	Area > 50 μm (mm^2)	(%)	Area > 50 μm (mm^2)	(%)
Fisher's gait	Cylindrical	604	8	674	9	695	9
	Tapered	642	8	740	10	771	10
	Anatomic	633	9	653	9	–	–
Duda's stair-climbing	Cylindrical	647	9	875	12	951	13
	Tapered	757	10	893	12	925	12
	Anatomic	562	8	584	8	–	–

though the survival rate dropped due to the wear of the polyethylene cup. Others [23] reported a survival rate of 97 % at 7–10 years, but again substantial polyethylene wear was observed in a significant percentage of the acetabular cups. It may not be appropriate, however, to compare these FE results with the follow-up studies of the actual ABG as the over-reaming of the distal part was not modelled. Other anatomical designs, which do not use an over-reaming technique, are the Profile and the APR-II. A follow-up study of the Profile at 10 years showed excellent clinical and radiographic results, with no stems revised for aseptic loosening and thigh pain was found in only 10 % of hips [7]. The APR-II stem has also been found to be very successful [24]. Out of 99 total hip arthroplasties followed for up to 4 years, 100 % had proximal bone ingrowth fixation and no patient reporting thigh pain after 3 years. Distal cortical hypertrophy associated with tip fixation occurred in 49 %, whereas proximal stress-shielding was present in 43 % of hips.

An in vitro experimental comparison study between a curved anatomical stem and a straight stem [25] found that at low angles of flexion, the curved and straight stems demonstrated similar patterns of motion. However, at high torsional moments such as the one observed during stair-climbing, the curved stem was found to be more stable than the straight stem. In another experiment comparing between straight and asymmetrical hip stems with 1,000 N of load applied to the prosthetic head, the axial micromotions were found to be small for both stems, with the symmetrical stem having the least axial micromotion—an average of 6 μm—compared to the asymmetrical stem with an average of 19 μm [26]. However, a torsional loading test showed that the straight stem had about ten times greater relative rotational motion than did the other two stems. Another study [27] showed that straight and curved stems performed similarly in terms of micromotion during single leg stance and at low loads during stair-climbing. When large torsional moments (22 Nm) were applied, the straight stem produced 2–4 times more micromotion than the curved stems. All these three papers reported that straight stems produced more micromotion than curved anatomic stems during stair-climbing. The FE results, however, did not show a significant difference between the two designs at this physiological loading. The reason could be that in the FE models, a perfect fit was created with an interference of 0.1 mm across the surface of the stem. The FE results showed that the straight stem had 9 % surface area in excess of the threshold limit of 50 μm interface micromotion compared to the anatomic with 8 %. However, when the bone elements were adjusted to simulate bone loss, the area for the straight stem increased to 12 %, but the anatomic design was maintained at 8 %. This showed that straight stems were more susceptible to micromotion during stair-climbing when a perfect fit at the interface was not achieved.

The Alloclassic stem, which represented the tapered group, is also a successful design with a survival rate of 99.3 to 100 % between 5 and 11 years [28–30]. In one of the follow-up reports of the Alloclassic [31], 98 % of the hips were rated good or excellent clinically at a median of 4 years. No stem was classified as definitely loose and no hips required revision. There was also no incidence of femoral osteolysis. Another study [32] reported that only 3 out of 133 stems subsided 2–5 mm and one subsided 5–10 mm within the first year, but no progressive subsidence could be detected beyond this period. Another [33] reported excellent results at 8 years for the Alloclassic with 83 % showing no radiolucency and 17 % showing radiolucency only proximally. A retrieval study of the Alloclassic [34] found that extensive bone-to-prosthesis apposition occurred at the interface along the stem between 6 weeks and 60 months. The mean appositional bone index was 48 %. Other tapered designs have also been shown to have excellent results at 5 years [35] and 10 years [4, 36, 37].

The models representing the three groups have been modified as much as possible so that a proper comparison could be made between them. However, the lateral flare feature of the Alloclassic was not removed because of the difficulty in redesigning the prosthesis in three-dimensions. The lateral flare is a proximal lateral expansion, which is designed to engage the lateral cortex of the femur in the metaphysis, allowing for a much broader base of support in this area. This allows a more concentric loading in the proximal femur and relieves distal stress transfer. It has been reported that this feature provides extra initial stability in cementless hip stems [38]. Another study [39] reported that prostheses with a lateral flare, such as the Alloclassic, have better rotational stability compared to the Schenker prosthesis, a stem without a lateral flare. The micromotion algorithm could only display resultant interface relative motion and could not separate this into axial and rotational components. As such the superior rotational micromotion of the Alloclassic could not be measured. The effect of a lateral flare feature will be analysed in Sect. 3.4 when proximal and distal fixation designs are discussed.

Rotational stability under torsional loads has been reported by Gortz et al. [40]. Four hip stems were measured experimentally, two of which belonged to the tapered group (the Alloclassic and the CLS), one to the anatomical group (the ABG), and one belonged to the straight cylindrical group (the S-ROM). They found that the relative rotational motion for the anatomical ABG and the tapered CLS was larger distally than proximally, whereas the Alloclassic, which is in the same group as the CLS, showed the opposite—larger proximally than distally. Though their findings could not be related to the FE results here, due to the limitation of the micromotion algorithm, it showed that the overall geometry alone could not totally account for the variable stability of cementless hip stems. Different concepts of fixation, namely proximal and distal fixation, also play a role in the stability of femoral components.

In conclusion, this section showed the results of an FE micromotion study that proved the stability of three categories of hip stems—the cylindrical, the tapered and the anatomical designs. Specific features of these implants were removed as much as possible so that a proper comparison between the three categories could be made. One of the limitations of this study was the inability to separate the resultant micromotion into axial and rotational components. It is therefore not clear, for example, which type of hip stem was better in terms of sustaining torsional loading. Despite these limitations, this study confirmed the stability of these types of hip stem designs.

3.2 The Material Stiffness

One of the design factors that must be taken into consideration is the type of material used, where a trade off between load transfer and implant stability is the key issue [41]. The earlier generation of cementless stems were too stiff compared to the bone—about 10 times too stiff. A stiff material will make load transfer inefficient, and eventually cause 'stress-shielding'—an adverse bone remodelling phenomenon where bone is resorbed in areas where it is not loaded to physiological levels. This stiffness mismatch between the stem and the bone causes loss of proximal cancellous bone and thickening of distal cortical bone. It was, and still is, one of the major problems in hip arthroplasties, both cemented and cementless.

The problem of load transfer from stiff implants to bone led to the development of low stiffness stems, sometimes called 'isoelastic' stems. The aim of isoelasticity was to deform the implant and the bone as one unit, thus maintaining the bone structure better. In terms of maintaining bone stock, compliant stems have been shown to be better than stiff stems. An in vivo study [42] on canine models showed that reduced stem stiffness enhanced proximal load transfer, thus reducing proximal bone loss. In another study of 14 patients, where 6 patients had isoelastic implants, their overall Bone Mineral Density (BMD)—a parameter that measures bone quality—increased by a mean of 12.6 %. For those with a relatively stiff titanium implant, BMD decreased by a mean of 27 % after 12 months [43]. An FE bone remodelling study comparing the effects of the modulus of elasticity of various stems also revealed that low stiffness material such as a CFRP composite reduced stress-shielding in the proximal bone by approximately one-half in comparison to the titanium alloy [44].

Though there seems to be an advantage of using isoelastic stems, several authors reported that the use of these stems caused high rate of aseptic loosening. The RM prosthesis and the Morscher prosthesis were two of the earliest isoelastic cementless stems. There was a high rate of aseptic loosening at follow-up period of 9 years [45]. A later generation of isoelastic stems such as the prototype carbon fibre-reinforced composite also suffered a similar fate [46]. The authors reported macroscopic aseptic loosening and fibrous interface fixation for 92 % of carbon fibre hip prosthesis at 6 years.

As far as the author is aware, FE analyses on the effects of material stiffness on micromotion have only been done using a simplified cylindrical model [47] or a simplified 2D model [48]. In this section, 3D FE analyses were conducted on three different material properties—cobalt chromium (CoCr), titanium alloy (TiAl) and composite material—to compare the micromotion between them under physiological loading.

3.2.1 Finite Element Modelling

Three AML stems prepared in Chap. 2 were analysed by assigning three different elastic moduli representing cobalt chromium (200 GPa), titanium alloy (110 GPa) and composite material (20 GPa). All stems were loaded using Fisher's walking loadcase and the results of interface micromotion were displayed.

3.2.2 Biomechanical Effect of Different Mechanical Properties

Surface areas of bone with more than 50 μm of micromotion were removed as described before to simulate interfacial bone loss. Further iterations were performed on each model with the same load cycle, and the new set of results were plotted and analysed.

Figure 3.8 shows that interface micromotion increased as the stiffness was reduced, with the feasible area for bone ingrowth reduced from 91 % for the cobalt

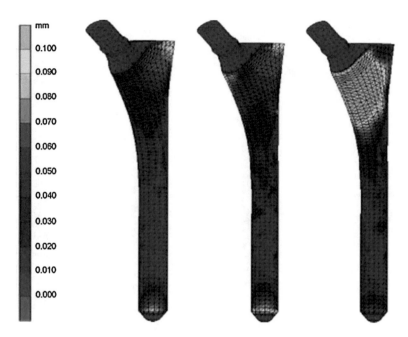

Fig. 3.8 Micromotion results for the AML stem made of CoCr (*left*), TiAl (*middle*) and composite (*right*) after 1st iteration

chromium to only 69 % for the composite. The area that was affected most by excessive micromotions was the proximal part, thus reducing the effectiveness of the implant in terms of maintaining stability. All load transfer then occurred in the distal part, relying on stiff cortical contact in that area.

When the unfeasible areas of bone ingrowth were removed and the analyses were repeated, the areas with excessive micromotion increased for all implants (Fig. 3.9). However, the most compliant material extended the most among the three stems (Figs. 3.10, 3.11 and Table 3.3). Cobalt chromium was the most stable with only an increase of 1 %, and composite was the worst with more than half of the stem, mostly proximal, had excessive micromotion beyond the threshold limit of bone ingrowth.

The flexible stems caused more micromotion than the stiff ones. This is in agreement with the experimental findings of Otani et al. [49]. They conducted micromotion experiments on cementless stems with three different material properties—carbon composite (E = 18.6 GPa), titanium alloy (100 GPa) and stainless steel (200 GPa). They found that the carbon composite stem produced significantly larger micromotion proximally and significantly smaller micromotion distally than in the two metals. They concluded that proximal stress transfer may be improved by a flexible stem, but they raised the possibility of increased micromotion in the proximal area. They have also suggested that improved proximal fixation may be necessary to achieve clinical success with flexible composite femoral components.

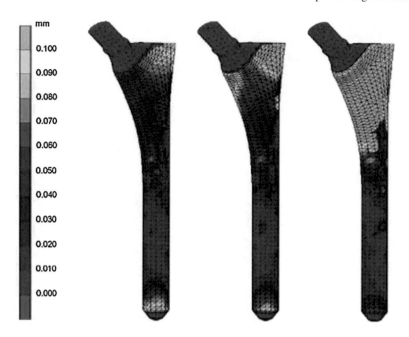

Fig. 3.9 Micromotion results for the AML stem made of CoCr (*left*), TiAl (*middle*) and composite (*right*) after final iteration

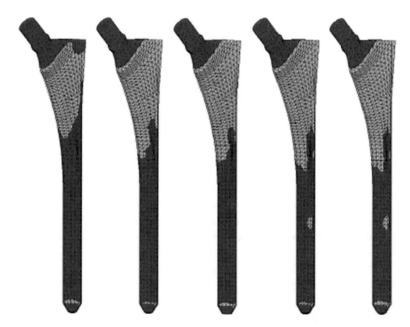

Fig. 3.10 Progression of surface area (*grey colour*) unfeasible for bone growth for the composite stem (iteration 1 to 5)

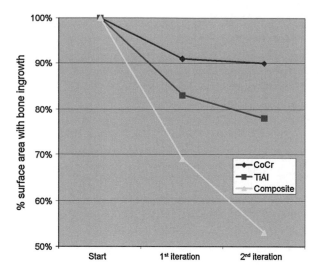

Fig. 3.11 The reduction in surface area with less than 50 μm for the CoCr, TiAl and composite stems

Table 3.3 Amount of surface area with more than 50 μm of micromotion and its percentage area (the total surface area of the stem is 8,976 mm²) up to the 4th iteration

	First iteration		Second iteration		Third iteration		Fourth iteration	
	Area > 50 μm (mm²)	(%)	Area > 50 μm (mm²)	(%)	Area > 50 μm (mm²)	(%)	Area > 50 μm (mm²)	(%)
CoCr	769	9	890	10	933	10	–	–
TiAl	1,542	17	1,955	22	2,037	23	2,105	23
Composite	2,757	31	4,204	47	4,481	50	4,546	51

There are numerous other studies comparing implants with different stiffnesses that showed significant increase in proximal micromotion for a flexible implant compared to titanium alloy or steel implants [50–52]. A pilot study on the goat [53], also revealed that stiff implants showed favourable initial interface micromotion for bone ingrowth.

These experimental observations were further supported by numerical analyses [47]. Using a simplified cylindrical model, the authors found that the interface micromotion reduced by about half for Co-Cr–Mo alloy compared to an implant with a stiffness ten times lower. Another finite element study on the effects of material stiffness used 2D model of an implanted prosthesis [48]. The authors found that a flexible stem generated motions about three to four times larger proximally than those of a stiff stem. These published results together with our findings showed that even if flexible implants could optimise load transfer, their flexibility could eliminate the environment that is needed for osseointegration.

Whilst rigid stems caused bone loss due to stiffness mismatch, compliant stems caused problems in stability because the implant is easier to displace and rotate than its stiffer counterpart [49, 54]. As a compromise between the effect of 'stress-shielding' of a stiff stem and the effect of 'excessive micromotion' of flexible stems, most hip stems nowadays are made of titanium alloy. Apart from having high structural strength and being relatively bio-inert, it has a stiffness value of around 100 GPa—still larger than the stiffness of bone, but less stiff than implants made of steel or cobalt chromium.

Early examples of flexible stems were the RM prosthesis [55, 56] and the Morscher prosthesis [45]. Both implants showed promising early results but a high rate of aseptic loosening in long-term follow-up. Femoral components were found to be extremely loose and easily removable, with a thick, shiny fibrous membrane at the interface—no evidence of any bone ingrowth whatsoever [57]. Too high flexibility in the proximal part of the prosthesis was blamed for bone resorption and implant loosening [45]. Adam et al. [46] reported macroscopic aseptic loosening and fibrous interface fixation for 92 % of carbon fibre-reinforced composite hip prostheses at 6 years. Another study comparing between isoelastic Butel stems and stiffer PCA stems showed that isoelastic stems showed a significantly higher rate of loosening (43 %) than the stiff stems at 4 years post-operatively [58]. The Butel stems, however, gave fewer signs of stress-shielding radiologically.

Despite the above reports of hip stems loosening with the use of compliant material, a recent short-term (2 years) follow-up study [59] reported an encouraging result of the isoelastic Epoch stems. The migration of the Epoch was not found to be statistically significant than the relatively stiffer Anatomic stems. Both stems stabilised with migration below 100 µm at 2 years with the Anatomic showing less migration than the Epoch. None of the stems were revised at the time of follow-up and the Epoch showed significantly reduced bone loss in Gruen zones 1, 2, 6 and 7 at 2 years compared to the Anatomic. The Epoch isoelastic stem seemed to show better short-term stability than the early generation of isoelastic stems such as the RM, the Morscher and the Butel. This could be attributed to other design aspects of the implant such as the overall geometry of the stem and the surface finish. The RM prosthesis, for example, was not designed to fill the canal whilst the Butel prosthesis had a smooth surface. The Epoch on the other hand is a 'fit and fill' design and has porous coating throughout the length of the stem. The findings from numerical simulation also showed that even though a compliant material produced larger micromotion than stiffer implants, with a 'fit and fill' design, the distal half of the stem was as stable as its stiffer counterpart.

This section analysed the effect of material stiffness on interface micromotion. Our analyses showed that micromotion increased as the stiffness was reduced. Our results were in agreement with other published reports with regard to material stiffness, whether FE, experimental or follow-up studies. In this study, not just comparative micromotion results were presented, but stability was also predicted for the three different hip stems' elastic moduli by simulating interfacial bone loss. The results showed that even though flexible stem produced larger micromotion and significant proximal interface bone loss, it should still be stable if tight fit was achieved distally.

3.3 The Effect of Stem Length

The study of stem length is particularly relevant for implants that are to be used in revision surgery. Due to loss of bone stock mostly in the proximal femur after a failed primary arthroplasty, surgeons have to rely on the cortical bone distally. As such, revision hip stems are normally longer than their primary counterparts in order to achieve proper stability through distal fixation. The optimum stem length is still a topic of discussion and debate in revision surgery [60].

There have been few studies on the optimum length of hip stems for primary arthroplasty. These looked into the effect of stem length for a specific type of hip prosthesis where the distal part of the stem was thinner than the medullary canal. Since the distal part of the stem did not fit and fill the canal, it may be hypothesized that the distal portion of the stem had no mechanical function, thus the study of stem length. One of these studies [61] compared three stem lengths of the Freeman hip stem - the full length of 172 mm, an intermediate length of 132 mm and a short stem of 92 mm. An FE model of each of them was created and analysed, and the actual prostheses with different stem lengths were tested experimentally. They found that increasing the length of the stem resulted in a sharp increase in the level of compressive stresses laterally at the tip. For the short stem, there was an increase in the proximal stresses medially while still having similar lateral stresses distally. All femora with short stem components also failed at loads between 800 and 1200 N, whereas no femora with longer stems failed. They concluded that a suitable length of stem was the intermediate one. Another FE study [62], on the other hand, showed that reducing stem length so as to effectively remove it did not increase failure probability, and it did not reduce stress-shielding either. They concluded that reducing the stem length to the metaphyseal area was not advantageous. These two studies, however, did not mention if stability was compromised by having a short stem.

Even though the study of stem length is mainly popular in revision surgery, it is also important for primary hip arthroplasty. If we could use as short a stem as possible initially, then should the need for revision surgery arise, the loss of bone stock would be much less than if we were to use a longer stem. As the two papers described above have already looked at the stresses at various stem lengths, the aim of this study is to investigate, using finite element analysis, the effect of stem lengths on interface micromotion.

Before starting to analyse the effects of stem length, it is important to note that this study could also be related to the study between proximal and distal fixation design, which will be discussed in the next section of this chapter. In the concept of proximal fixation, if the implant is loaded proximally, then there may be no need of a long stem. However, design characteristics of a proximally-fixed design are more than just having a short stem, and as such a separate study of proximal versus distal fixation stems is presented in a separate section.

3.3.1 Finite Element Modelling

Four models of the AML stems were adjusted to a different length—the standard 175 mm and three shorter ones—135 mm (medium length), 95 mm (short length) and 74 mm (very short). The length of the shortest one was determined using the CT dataset so that minimum cortical contact was obtained. All stems were loaded in Fisher's physiological walking loadcase and the micromotion results were plotted.

3.3.2 Biomechanical Assessment of Different Stem Lengths

Areas with more than 50 μm of micromotion were then removed to simulate interfacial bone loss and the analyses were repeated until a stable-state was achieved or failure occurred. Figure 3.12 shows that shortening the stem increased the interface micromotion, with a huge increase in micromotion between the third stem (short length) and the fourth stem (very short). For the shortest implant, further iterations were not conducted because 79 % of the stem had micromotion in excess of 50 μm. The results of the first three stems from the final iteration (Fig. 3.13) showed similar results—all stems showed an increase in interface micromotion with the short stem having the largest micromotion (Fig. 3.14).

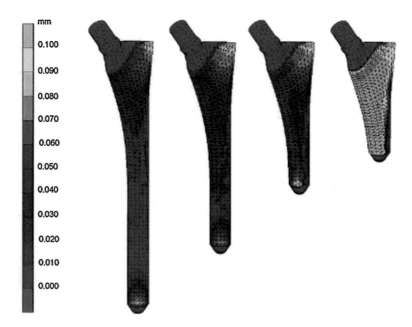

Fig. 3.12 Micromotion results for the AML at various stem length for the 1st iteration

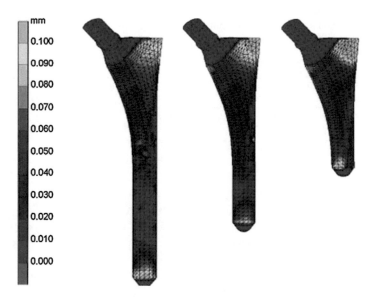

Fig. 3.13 Micromotion results for the AML at various stem length for the final iteration. The fourth model was not included as the surface was surrounded with micromotion in excess of the chosen threshold limit

Fig. 3.14 The reduction in surface area with less than 50 μm of micromotion for the standard, medium and short stems

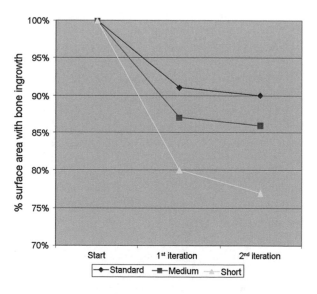

The surface area feasible for bone ingrowth reduced as the length of the stem was reduced, both in terms of quantity and relative percentage (Table 3.4). However, in terms of overall stability, the first three stem lengths were found to be stable, with more than 75 % of their surface areas found to be feasible for bone ingrowth.

Table 3.4 The amount of surface area with more than 50 μm of micromotion and its percentage area for the standard (8,976 mm²), medium (7,284 mm²), short (5,592 mm²) and very short stems (4,536 mm²)

	First iteration		Second iteration		Third iteration		Fourth iteration	
	Area > 50 μm (mm²)	(%)	Area > 50 μm (mm²)	(%)	Area > 50 μm (mm²)	(%)	Area > 50 μm (mm²)	(%)
Standard	769	9	890	10	933	10	–	–
Medium	938	13	993	14	1,040	14	–	–
Short	1,114	20	1,275	23	1,342	24	1,364	24
Very short	3,565	79	–	–	–	–	–	–

The findings of the study on the stem length showed the importance of cortical contact in achieving stability. The length of the very short stem was chosen so that contact between the stem and cortical bone was minimal. The FE results showed that this stem had the largest micromotion of all.

Among the first three stem lengths, the short stem had micromotion larger and more widely spread. However, the results from the final iteration of simulated bone loss showed that this did not affect its stability. The only available study to compare these results with was the work of Sakai et al. [63], where 60 patients underwent hip replacement with custom-made femoral components of two different lengths. At the period of follow-up, there were no statistical differences clinically and radiographically between the 125 mm long stem and the 100 mm long stem. Apart from this comparative study, there is also a follow-up study of short-stem Mayo prosthesis [64]. The prosthesis has a double-wedged contour and fixed in place using 3-point proximal femur fixation. The follow-up period was 1 year and from out of 20 patients, one was revised for loosening while the rest were classified as satisfactory. It is surprising that a longer-term review has not been published, and it is certainly discouraging to see that a stem had loosened in 1 year. As there are not so many reports on short-stemmed prostheses, it is difficult to make a solid conclusion on their stability. However, the two reports above showed that as long as there was stiff cortical contact on the stem, primary stability was not compromised. This is confirmed by this FE study, that even though shorter stems have more and larger micromotion than longer stems, they should still be stable.

The main advantage of having a short-stemmed prosthesis over long-stemmed is the ease of converting to another implant should failure occur. Since less bone is removed, it provides a possibility of long-term compatibility. As far as interface micromotion is concerned, a short-stemmed prosthesis could be made more stable through other design features, such as a pronounced lateral flare that rests on the inferolateral part of the greater trochanter or by having proximal macrofeatures. This study showed that shorter stems could have the potential of becoming the next generation of cementless hip stems if their stability can be improved.

3.4 Proximal Versus Distal Fixation

As described at the start of this chapter, there are two main design philosophies of fixation of cementless hip stems—proximal fixation and distal fixation. Distal fixation is usually achieved through the press-fit technique where the distal part of the stem is fixed to the cortical diaphysis, one example of which is the AML. Proximal fixation, on the other hand, relies for its stability on the cancellous bone surrounding the proximal part, such as the ABG hip stem. Design features for both concepts of fixation can vary, such as having a straight or curved stem, rectangular or cylindrical shape, collared or non-collared, types of materials used, length of stem and types and extent of coating.

Proximal or distal fixations are both design concepts for reducing bone loss, especially in the proximal area by improving the load transfer [65]. Which design concept is better remains a controversial issue. Advocates of the distal fixation concept argue that strong primary stability is crucial and it is achievable through strong and reliable cortical contact distally. Once the implant is stably fixed, bone ingrowth can then take place throughout the stem. Advocates of the proximal fixation concept argue that bone resorption in the proximal area can be reduced if the forces from the hip joint are transmitted to the most proximal part of the femur. An additional rationale for using a metaphyseal filling, proximal fixation design is to preserve the endosteum of the diaphysis for later surgery, should a revision be required in the future.

The design of the distal and proximal fixation varies. The proximal fixation design was attributed to anatomical stems such as the APR-II stem where the proximal implant geometry should fit and fill as much as possible of the proximal femur [65]. As such, the proximal part is widened in the antero-posterior and medio-lateral directions to allow a greater area of bone attachment proximally. However, the surgical technique of the APR-II stems employs a tight circumferential diaphyseal fit [66]. It is therefore uncertain if total proximal load transfer is achievable. Later generations of proximal fixation design, such as the ABG and the IPS, used distal over-reaming technique as standard surgical protocol. Theoretically this will allow more proximal loading than the diaphyseally fit technique used for the APR-II. Another advantage of distal over-reaming is that it significantly reduced thigh pain [22]. This over-reaming technique has also been tested on the distal fixation AML stem [8]. The authors reported that fractures were more likely to occur using the under-reaming technique, and therefore switched to over-reaming the distal endosteal bone. Both techniques were found to give successful outcomes.

A tapered design can be either distally or proximally fixed because stability is achieved by wedging the implant into the femur. The Alloclassic, for example, achieves a wedge-fit in the diaphysis and therefore is categorised as a distal fixation design. The CLS, on the other hand, has a tri-tapered design with a rectangular cross-section and a narrow stem tip. A wedge-fit is achieved in the metaphyseal area and therefore it is considered to be a proximal fixation design.

In this section, FE analyses will be conducted to compare interface micromotion, and therefore the stability, of two design concepts for cementless fixation—the proximal fixation design and the distal fixation design.

3.4.1 Finite Element Modelling

In order to make a proper and general comparison between proximal and distal fixation stems, a design was chosen to represent both concepts so that the minimum changes possible can be made while maintaining the main features of the design concept. The AML was chosen as the distal fixation design. To create a proximal design from the AML, the stem was shortened by a half, and the proximal part was enlarged in the medio-lateral and antero-posterior directions in order to help it to engage the proximal cortical shell (Fig. 3.15). The medio-lateral size was enlarged by having a lateral enlargement that filled the infero-lateral part of the greater trochanter. The distal part of the stem was shortened by 80 mm, and the endosteal cavity was over-reamed by 28 mm distally, making the effective length of the stem 74 mm. This was done because stems intended for proximal load transfer usually have the distal cavity over-reamed, therefore the distal stem (if perfectly aligned) plays no part in load transfer. Both designs were loaded in walking and stair-climbing modes.

3.4.2 Biomechanical Effect of Different Stem Fixations

The proximal fixation design had more micromotion than the distal fixation design in both walking and stair-climbing as depicted in Fig. 3.16. The distribution of micromotion was also different between the two designs. The distal fixation model

Fig. 3.15 The distal fixation design (*left*) and the proximal fixation design showing enlargement in medio-lateral and antero-posterior directions (*right*)

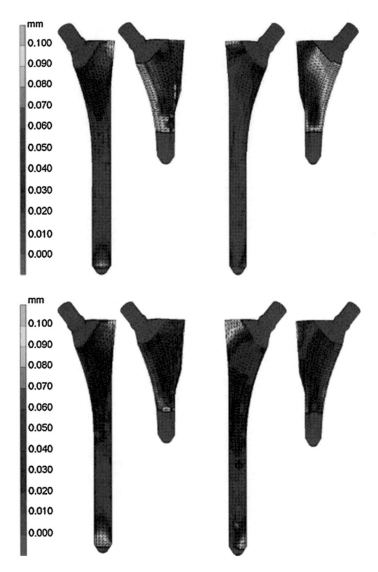

Fig. 3.16 Contour plots of micromotion for a distal fixation and a proximal fixation design using Fisher's gait cycle (*top*) and Duda's stair-climbing (*bottom*) after 1st iteration. Posterior side on the left, anterior side on the right

had micromotion largely in the lateral area of the proximal part and around the anterior tip of the stem. The proximal design, on the other hand, had micromotion concentrated more on the medial part of the stem.

Table 3.5 shows that unfeasible areas for bone ingrowth were larger in the proximal design by 2–6 times compared to the distal design. For the proximal fixation design, the stem seemed to be more unstable in walking, with surface

Table 3.5 Surface area more than 50 μm of micromotion for the distal fixation design (8,976 mm^2) and the proximal fixation design (5,370 mm^2)

		First iteration		Second iteration	
		Area > 50 μm (mm^2)	(%)	Area > 50 μm (mm^2)	(%)
Distal fixation	Fisher's gait	769	9	890	10
	Duda's stair-climbing	999	11	1,113	12
Proximal fixation	Fisher's gait	2,899	54	5,105	95
	Duda's stair-climbing	1,257	23	4,274	80

areas exceeding the threshold limit more than double compared to surface areas during stair-climbing. However, when interfacial bone loss was simulated, the proximal fixation design failed in both physiological loadings (Figs. 3.17, 3.18 and 3.19).

The distal fixation design was more stable than proximal fixation design by up to six times based on 50 μm threshold limit for bone ingrowth. This could be attributed to the larger surface area in contact with the cortical bone for the distal design. A follow-up study [67] comparing between the straight stem distal fixation Bi-Metric stem with the distally over-reamed proximal fixation ABG stem, showed that there was substantially more cortical contact in the diaphyseal area for the distal fixation design as expected, and that subsidence of more than 2 mm was more frequent in the proximal fixation ABG, though the stability was not compromised. Both hip stem designs, however, were found to be clinically excellent at 5 years. There was a tight diaphyseal fit that produced excellent stability for the distal fixation design, but this meant that stress was transferred in this region and as a result increased stress-shielding in the proximal metaphyseal compared to the proximal fixation ABG.

For the stability of proximal design in stair-climbing, this has also been confirmed by others, as mentioned in the previous section. Hips stems with a lateral flare feature in its design, has been proven to provide extra initial stability particularly the rotational stability compared to a stem without lateral flare [39]. Another study [68] also confirmed that a lateral flare contributed positively to a cementless stem design and that the stems could be made shorter than designs without a lateral flare. In their FE study, migration was found to be less for the lateral flare compared to a simple straight stem model. A follow-up study of the lateral flare system by the same authors also showed that trabecular bone attached to the lateral flare coated with HA. Another paper [38] reported that cementless stems with a lateral flare provided primary stability and produced a low subsidence—an average of 0.32 mm at 2 years.

An experimental study looking at the rotational stability of the two fixation designs has been reported [40]. Two proximally fixed stems, the ABG and the CLS, were found to have more rotational micromotion distally than proximally. The distal fixation Alloclassic, however, was found to produce more rotational micromotion proximally than distally. The rotational micromotion in the proximal

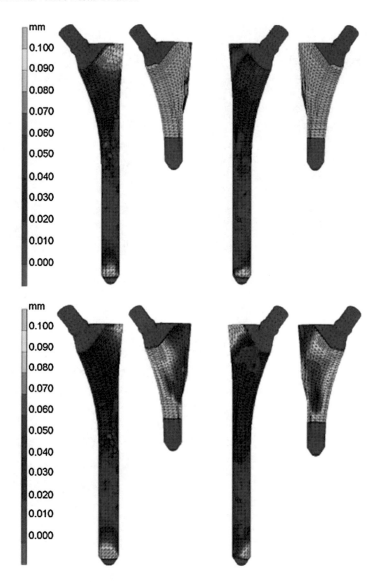

Fig. 3.17 Contour plots of micromotion for a distal fixation and a proximal fixation design using Fisher's gait cycle (*top*) and Duda's stair-climbing (*bottom*) after final iteration. Posterior side on the left, anterior side on the right

area was larger for the Alloclassic by two times compared to the CLS and almost three times compared to the ABG. Better rotational stability for the two proximal fixation designs could be due to the proximal macrofeatures on the anterior and posterior sides of both hip stems. The CLS had tapered fins macrofeature and the ABG had semi-circular indentations macrofeature.

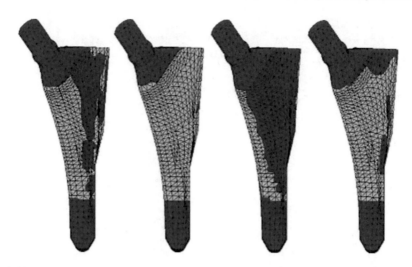

Fig. 3.18 Progression of surface area (*grey colour*) unfeasible for bone growth for the proximal fixation stem (iteration 1 and 2) for the walking load (*left two*) and stair-climbing load (*right two*)

Fig. 3.19 The reduction in surface area with less than 50 μm of micromotion for the distal and proximal fixation designs using Fisher's gait cycle and Duda's stair-climbing loads

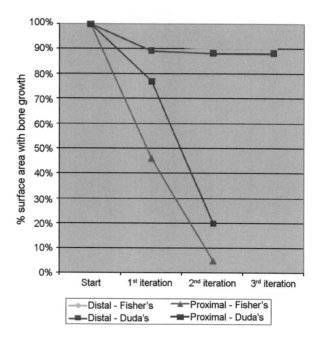

Another in vitro experiment looked at four different cementless stems [69]. One of the tested stems was the CLS and another one was the Zweymuller which has a tapered design similar to the distally fixed Alloclassic. The others were the Muller 85 and the anatomical PCA. They found that the CLS had the largest mean interface micromotion under load simulating single leg stance—up to 10 times more motion than the most stable stem (the anatomical PCA). Another study reported the migration of the proximally-fixed CLS stems [70]. The mean femoral stem migration was 2 mm at 2 years and 3.66 mm at 7 years. Although these reports showed more migration and micromotion for a proximal fixation design, follow-up studies of the ABG and the CLS showed that these prostheses were successful [71, 72].

The findings from FE analysis showed that the proximal fixation design, though less stable than the distal fixation design, has some potential. As discussed in the previous section, the advantage of having a short-stemmed prosthesis is its long-term compatibility. Should the need for revision arise, less problems will be encountered as he cortical bone distally is available for fixation. However, the results from the previous section showed that full proximal load transfer by having a very short stem caused large interface micromotion and instability. In this section, the micromotion of short-stemmed prosthesis was reduced by having a proximal enlargement in the antero-posterior and medio-lateral direction. The stem failed in both physiological loadings when bone loss was simulated because most of the excessive micromotion concentrated on the medial area—stability was compromised because contact in the medial area was lost. The relative lack of stability for the proximal fixation design in this section could be improved further by having an anatomic design and introducing macrofeatures in the proximal part.

References

1. Healy WL (2002) Hip implant selection for total hip arthroplasty in elderly patients. Clin Orthop Relat Res 405:54–64
2. Mallory TH, Lombardi AV Jr, Leith JR, Fujita H, Hartman JF, Capps SG, Kefauver CA, Adams JB, Vorys GC (2001) Minimal 10-year results of a tapered cementless femoral component in total hip arthroplasty. J Arthroplasty 16(8 Suppl 1):49–54
3. Bourne RB, Rorabeck CH (1998) A critical look at cementless stems. Taper designs and when to use alternatives. Clin Orthop Relat Res 355:212–223
4. Reitman RD, Emerson R, Higgins L, Head W (2003) Thirteen year results of total hip arthroplasty using a tapered titanium femoral component inserted without cement in patients with type C bone. J Arthroplasty 18(Suppl 0):116–121
5. Parvizi J, Keisu KS, Hozack WJ, Sharkey PF, Rothman RH (2004) Primary total hip arthroplasty with an uncemented femoral component: a long-term study of the Taperloc stem. J Arthroplasty 19(2):151–156
6. Howard JL, Hui AJ, Bourne RB, McCalden RW, MacDonald SJ, Rorabeck CH (2004) A quantitative analysis of bone support comparing cementless tapered and distal fixation total hip replacements. J Arthroplasty 19(3):266–273
7. Kim YH, Oh SH, Kim JS (2003) Primary total hip arthroplasty with a second-generation cementless total hip prosthesis in patients younger than fifty years of age. J Bone Joint Surg Am 85-A(1):109–114
8. Kim YH, Kim VE (1994) Cementless porous-coated anatomic medullary locking total hip prostheses. J Arthroplasty 9(3):243–252
9. Engh CA (1998) Mini-symposium: total hip replacement—(ii) Distal porous coating yields optimal fixation. Curr Orthopaed 12(4):232–238
10. Engh CA, O'Connor D, Jasty M, McGovern TF, Bobyn JD, Harris WH (1992) Quantification of implant micromotion, strain shielding, and bone resorption with porous-coated anatomic medullary locking femoral prostheses. Clin Orthop Relat Res 285:13–29

11. Chess DG, Grainger RW, Phillips T, Zarzour ZD, Sheppard BR (1996) The cementless ana-
    tomic medullary locking femoral component: an independent clinical and radiographic
    assessment. Can J Surg 39(5):389–392
12. Engh CA, Hooten JPJ, Zettl-Schaffer K, Ghaffarpour M, McGovern TF, Macalino GE, Zicat
    BA (1994) Porous-coated total hip replacement. Clin Orthop Relat Res 298:89–96
13. Nercessian OA, Wu WH, Sarkissian H (2001) Clinical and radiographic results of cementless
    AML total hip arthroplasty in young patients. J Arthroplasty 16(3):312–316
14. Kronick JL, Barba ML, Paprosky WG (1997) Extensively coated femoral components in
    young patients. Clin Orthop Relat Res 344:263–274
15. Woolson ST, Adler NS (2002) The effect of partial or full weight bearing ambulation after
    cementless total hip arthroplasty. J Arthroplasty 17(7):820–825
16. Engh CA, Bobyn JD, Glassman AH (1987) Porous-coated hip replacement. The factors gov-
    erning bone ingrowth, stress shielding, and clinical results. J Bone Joint Surg Br 69(1):45–55
17. Whiteside LA, White SE, Engh CA, Head W (1993) Mechanical evaluation of cadaver
    retrieval specimens of cementless bone-ingrown total hip arthroplasty femoral components.
    The Journal of Arthroplasty 8(2):147–155
18. Sugiyama H, Whiteside LA, Engh CA, Otani T (1994) Late mechanical stability of the proxi-
    mal coated AML prosthesis. Orthopedics 17(7):583–588
19. Rossi P, Sibelli P, Fumero S, Crua E (1995) Short-term results of hydroxyapatite-coated pri-
    mary total hip arthroplasty. Clin Orthop Relat Res 310:98–102
20. Tonino AJ, Romanini L, Rossi P, Borroni M, Greco F, Garcia-Araujo C, Garcia-Dihinx L,
    Murcia-Mazon A, Hein W, Anderson J (1995) Hydroxyapatite-coated hip prostheses. Early
    results from an international study. Clin Orthop Relat Res 312:211–225
21. Garcia Araujo C, Fernandez Gonzalez J, Tonino A (1998) Rheumatoid arthritis and
    hydroxyapatite-coated hip prostheses: five-year results. Int ABG Study Group J Arthroplasty
    13(6):660–667
22. Giannikas KA, Din R, Sadiq S, Dunningham TH (2002) Medium-term results of the ABG
    total hip arthroplasty in young patients. J Arthroplasty 17(2):184–188
23. Herrera A, Canales V, Anderson J, Garcia-Araujo C, Murcia-Mazon A, Tonino AJ (2004)
    Seven to 10 years followup of an anatomic hip prosthesis: an international study. Clin Orthop
    Relat Res 423:129–137
24. Kang JS, Dorr LD, Wan Z (2000) The effect of diaphyseal biologic fixation on clinical results
    and fixation of the APR-II stem. J Arthroplasty 15(6):730–735
25. Berzins A, Sumner DR, Andriacchi TP, Galante JO (1993) Stem curvature and load angle
    influence the initial relative bone-implant motion of cementless femoral stems. J Orthop Res
    11(5):758–769
26. Hua J, Walker PS (1994) Relative motion of hip stems under load. An in vitro study of
    symmetrical, asymmetrical, and custom asymmetrical designs. J Bone Joint Surg Am
    76(1):95–103
27. Callaghan JJ, Fulghum CS, Glisson RR, Stranne SK (1992) The effect of femoral stem
    geometry on interface motion in uncemented porous-coated total hip prostheses. Comparison
    of straight-stem and curved-stem designs. J Bone Joint Surg Am 74(6):839–848
28. Delaunay C, Cazeau C, Kapandji AI (1998) Cementless primary total hip replacement: Four
    to eight year results with the Zweymuller-Alloclassic (R) prosthesis. Int Orthop 22:1–5
29. Pieringer H, Auersperg V, Grießler W, Böhler N (2003) Long-term results with the cement-
    less alloclassic brand hip arthroplasty system. J Arthroplasty 18(3):321–328
30. Pieringer H, Labek G, Auersperg V, Bohler N (2003) Cementless total hip arthroplasty in
    patients older than 80 years of age. J Bone Joint Surg Br 85(5):641–645
31. Huo MH, Martin RP, Zatorski LE, Keggi KJ (1995) Total hip arthroplasty using the
    Zweymuller stem implanted without cement: a prospective study of consecutive patients with
    minimum 3-year follow-up period. J Arthroplasty 10(6):793–799
32. Delaunay C, Fo Bonnomet, North J, Jobard D, Cazeau C, J-Fo Kempf (2001) Grit-blasted
    titanium femoral stem in cementless primary total hip arthroplasty: A 5- to 10-year multi-
    center study. J Arthroplasty 16(1):47–54

33. Effenberger H, Ramsauer T, Bohm G, Hilzensauer G, Dorn U, Lintner F (2002) Successful hip arthroplasty using cementless titanium implants in rheumatoid arthritis. Arch Orthop Trauma Surg 122(2):80–87

34. Lester DK (1997) Cross-section radiographic analysis of 10 retrieved titanium alloy press-fit femoral endoprostheses. J Arthroplasty 12(8):930–937

35. Keisu KS, Orozco F, McCallum Iii JD, Bissett G, Hozack WJ, Sharkey PF, Rothman RH (2001) Cementless femoral fixation in the rheumatoid patient undergoing total hip arthroplasty: Minimum 5-year results. J Arthroplasty 16(4):415–421

36. Eingartner C, Volkmann R, Winter E, Maurer F, Sauer G, Weller S, Weise K (2000) Results of an uncemented straight femoral shaft prosthesis after 9 years of follow-up. J Arthroplasty 15(4):440–447

37. Park M-S, Choi B-W, Kim S-J, Park J-H (2003) Plasma spray-coated Ti femoral component for cementless total hip arthroplasty. J Arthroplasty 18(5):626–630

38. Leali A, Fetto J, Insler H, Elfenbein D (2002) The effect of a lateral flare feature on implant stability. Int Orthop 26(3):166–169

39. Effenberger H, Heiland A, Ramsauer T, Plitz W, Dorn U (2001) A model for assessing the rotational stability of uncemented femoral implants. Arch Orthop Trauma Surg 121(1–2):60–64

40. Gortz W, Nagerl UV, Nagerl H, Thomsen M (2002) Spatial micromovements of uncemented femoral components after torsional loads. J Biomech Eng 124(6):706–713

41. Maistrelli GL, Fornasier V, Binnington A, McKenzie K, Sessa V, Harrington I (1991) Effect of stem modulus in a total hip arthroplasty model. J Bone Joint Surg Br 73(1):43–46

42. Sumner DR, Turner TM, Igloria R, Urban RM, Galante JO (1998) Functional adaptation and ingrowth of bone vary as a function of hip implant stiffness. J Biomech 31(10):909–917

43. Ang KC, De Das S, Goh JC, Low SL, Bose K (1997) Periprosthetic bone remodelling after cementless total hip replacement. A prospective comparison of two different implant designs. J Bone Joint Surg Br 79(4):675–679

44. Cheal EJ, Spector M, Hayes WC (1992) Role of loads and prosthesis material properties on the mechanics of the proximal femur after total hip arthroplasty. J Orthop Res 10(3):405–422

45. Morscher EW, Dick W (1983) Cementless fixation of "isoelastic" hip endoprostheses manufactured from plastic materials. Clin Orthop Relat Res 176:77–87

46. Adam F, Hammer DS, Pfautsch S, Westermann K (2002) Early failure of a press-fit carbon fiber hip prosthesis with a smooth surface. J Arthroplasty 17(2):217–223

47. Rohlmann A, Cheal EJ, Hayes WC, Bergmann G (1988) A nonlinear finite element analysis of interface conditions in porous coated hip endoprostheses. J Biomech 21(7):605–611

48. Kuiper JH, Huiskes R (1996) Friction and stem stiffness affect dynamic interface motion in total hip replacement. J Orthopaed Res 14(1):36–43

49. Otani T, Whiteside LA, White SE, McCarthy DS (1993) Effects of femoral component material properties on cementless fixation in total hip arthroplasty: a comparison study between carbon composite, titanium alloy, and stainless steel. J Arthroplasty 8(1):67–74

50. Sumner DR, Turner TM, Urban RM, Galante JO (1991) The bone-biomaterial interface. Bone ingrowth into porous coatings attached to prosthesis of differing stiffness. University of Toronto Press, Ontario, Canada

51. Yildiz H, Chang F-K, Goodman S (1998) Composite hip prosthesis design. II. Simulation. J Biomed Mater Res 39(1):102–119

52. Yildiz H, Ha S-K, Chang F-K (1998) Composite hip prosthesis design. I. Analysis. J Biomed Mater Res 39(1):92–101

53. Buma P, van Loon PJM, Versleyen H, Weinans H, Slooff TJJH, de Groot K, Huiskes R (1997) Histological and biomechanical analysis of bone and interface reactions around hydroxyapatite-coated intramedullary implants of different stiffness: a pilot study on the goat. Biomaterials 18(18):1251–1260

54. Anderson Engh Jr C, Sychters C, Charles E Sr (1999) Factors affecting femoral bone remodeling after cementless total hip arthroplasty. J Arthroplasty 14(5):637–644

55. Morscher E, Bombelli R, Schenk R, Mathys R (1981) The treatment of femoral neck fractures with an isoelastic endoprosthesis implanted without bone cement. Arch Orthop Trauma Surg 98(2):93–100

56. Niinimaki T, Puranen J, Jalovaara P (1994) Total hip arthroplasty using isoelastic femoral stems. A seven- to nine-year follow-up in 108 patients. J Bone Joint Surg Br 76(3):413–418

57. Jakim I, Barlin C, Sweet MB (1988) RM isoelastic total hip arthroplasty. A review of 34 cases. J Arthroplasty 3(3):191–199

58. Jacobsson SA, Djerf K, Gillquist J, Hammerby S, Ivarsson I (1993) A prospective comparison of Butel and PCA hip arthroplasty. J Bone Joint Surg Br 75(4):624–629

59. Karrholm J, Anderberg C, Snorrason F, Thanner J, Langeland N, Malchau H, Herberts P (2002) Evaluation of a femoral stem with reduced stiffness. A randomized study with use of radiostereometry and bone densitometry. J Bone Joint Surg Am 84-A(9):1651–1658

60. Mann KA, Ayers DC, Damron TA (1997) Effects of stem length on mechanics of the femoral hip component after cemented revision. J Orthopaed Res 15(1):62–68

61. Tanner KE, Yettram AL, Loeffler M, Goodier WD, Freeman MAR, Bonfield W (1995) Is stem length important in uncemented endoprostheses? Med Eng Phys 17(4):291–296

62. van Rietbergen B, Huiskes R (2001) Load transfer and stress shielding of the hydroxyapatite-ABG hip: a study of stem length and proximal fixation. J Arthroplasty 16(8, Suppl 1):55–63

63. Sakai T, Sugano N, Nishii T, Haraguchi K, Ochi T, Ohzono K (1999) Stem length and canal filling in uncemented custom-made total hip arthroplasty. Int Orthop 23(4):219–223

64. Morrey BF (1989) Short-stemmed uncemented femoral component for primary hip arthroplasty. Clin Orthop Relat Res 249:169–175

65. Longjohn DB, Dorr LD (1998) (iii) Proximal fixation stems yield optimal fixation. Curr Orthopaed 12(4):239–243

66. Dorr LD, Wan Z (1996) Comparative results of a distal modular sleeve, circumferential coating, and stiffness relief using the anatomic porous replacement II. J Arthroplasty 11(4):419–428

67. Laine HJ, Puolakka TJ, Moilanen T, Pajamaki KJ, Wirta J, Lehto MU (2000) The effects of cementless femoral stem shape and proximal surface texture on 'fit-and-fill' characteristics and on bone remodeling. Int Orthop 24(4):184–190

68. Walker PS, Culligan SG, Hua J, Muirhead-Allwood SK, Bentley G (1999) The effect of a lateral flare feature on uncemented hip stems. Hip Int 9(2):71–80

69. Schneider E, Kinast C, Eulenberger J, Wyder D, Eskilsson G, Perren SM (1989) A comparative study of the initial stability of cementless hip prostheses. Clin Orthop Relat Res 248:200–209

70. Davies MS, Parker BC, Ward DA, Hua J, Walker PS (1999) Migration of the uncemented CLS femoral component. Orthopedics 22(2):225–228

71. Rogers A, Kulkarni R, Downes EM (2003) The ABG hydroxyapatite-coated hip prosthesis: One hundred consecutive operations with average 6-year follow-up. J Arthroplasty 18(5):619–625

72. Schreiner U, Scheller G, Herbig J, Jani L (2001) Mid-term results of the cementless CLS stem. A 7- to 11-year follow-up study. Arch Orthop Trauma Surg 121(6):321–324

Chapter 4
Surgical and Pathological Parameters Affecting Micromotion

Abstract This chapter analysed two of the most important parameters affecting micromotion and therefore stability of hip implant. It has been accepted that interfacial gaps exist when broaching is used to prepare the bone bed for implantation. An experiment was carried out to quantify the gaps and analysis was performed to check on primary stability. No significant differences were observed between those with gaps and those without gaps. Other analyses were performed interfacial gaps set at potential locations of a particular implant. Similar results were observed where no significant differences were found in terms of primary stability. Undersizing and malalignment is another important error during surgery and is more prominent in young surgeons. The results show that undersizing and malalignment is more problematic for straight cylindrical stem compared to tapered stem. The effect of bone properties were also analysed where an extra care should be given to osteoporotic bone as the thicker implant diameter may not only caused potential fracture but also produced higher micromotion.

Keywords Interfacial gaps • Undersizing and malalignment • Osteoporotic • Surgical parameters • Bone pathology

4.1 Surgical Gaps

Implantation of cementless hip stems requires specific surgical procedures based on their design. However, the principles of implantation are similar; satisfactory femoral exposure, opening the femoral canal, broaching the proximal femur (either hand-broaching or power-reaming), irrigation of the femoral canal, and implanting the femoral stem to a position which is stable to bending, tilting, rotation, subsidence and pull-out force [1].

Gaps between implant and bone may occur especially when contemporary broaching techniques are used which are subject to surgical error. Gaps occur at various locations along the stem because the broach that prepares the bony bed passes

M. R. Abdul Kadir, *Computational Biomechanics of the Hip Joint*, SpringerBriefs in Computational Mechanics, DOI: 10.1007/978-3-642-38777-7_4,

Table 4.1 Comparison between traditional and surgical robot for femoral canal preparation [4]

	Traditional	Surgical robot
Cavity oversizing	22–40 %	0.4–0.7 %
Linear gap	1.0–3.5 mm	0.03–0.08 mm
Perimeter in contact with bone	21 %	96 %

outside the outline of the femoral component [2]. Noble et al. [3] have reported the difficulties of achieving interfacial contact due to the anatomical variations of the femur. Surface-to-surface contact between implant and bone was not achieved except at discrete areas of the interface. A study was conducted by Paul et al. [4] to compare the accuracy of traditional broaching and reaming for femoral canal preparation versus a surgical robot for milling the cavity. In their study they measured 'cavity oversizing', 'linear gap' and 'percentage of perimeter in contact with the bone' and the results showed that robotic milling produced a superior accuracy in terms of implant fit (Table 4.1). The study, however, fell short of analysing the effect of the inadequacies of using traditional techniques on primary stability.

The experimental findings of Paul et al. [4] have led to the development of an algorithm that can show interfacial contact and the existence of gaps at the bone-implant interface [5]. Though the paper only described the validation procedure of the algorithm, the examples presented in the paper showed that interfacial contact did not occur throughout the stem-bone interface.

Pazzaglia et al. [6] studied the effect of interfacial gaps on rats. Two groups of roughened titanium rods were examined—one with an interference fit and the other with a 0.3 mm diametrical gap at the interface. They found that there was no bone integration at the interface of the over-reamed canal, whether the surface of the rod was coated with hydroxyapatite or not. The interface was surrounded by a thick layer of fibrous tissue, with no bony connections with the endosteal surface.

To the author's knowledge, there are no reported studies that have looked into the effects of gaps on primary stability. In order to carry out this study, a technique of identifying interfacial gaps is proposed (the first method), where multiple CT scans of the specimen are required. In cases where specimens are not available, discussions are conducted with experienced orthopaedic surgeons to determine the areas where interfacial gaps are most likely to occur (the second method). The aim is to discover if these interfacial gaps created due to surgical error or otherwise, have any effects on the primary stability of cementless hip stems.

4.1.1 Finite Element Modelling: First Method

In order to study interfacial gaps created during broaching, two sets of images from two separate cadavers implanted with the Alloclassic hip stem were scanned using normal CT and PQCT. From these two sets of images interfacial gaps were found on both (Fig. 4.1). However, artefacts from the metallic implant made it difficult to determine

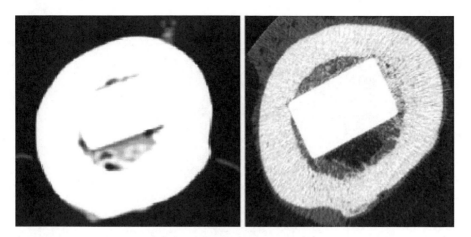

Fig. 4.1 Normal CT and PQCT from two different cadavers with Alloclassic hip implant showing interfacial gaps on the posterior side

the interfacial gaps accurately. In order to see more clearly the existence of gaps at the interface, a procedure is proposed as described in the following paragraphs.

A single dataset of implanted bone such as the one shown above is not sufficient to create an FE model to study interfacial gaps. The artefacts due to the metallic material of the implant caused several problems. Apart from the problem in analysing the gaps, the implant model created from the digitised image was also not appropriate to be used in FE analyses. Implant construction using this single dataset failed to create a perfectly shaped implant—they are coarse and slightly larger in size than the actual implant (Fig. 4.2—*right*). The model of the implant must therefore be created from the original 3D CAD model. The CT dataset of an implanted femur also cannot be used even if the implant model was constructed separately. This is because bone material properties assignment, as described in Chap. 2, will show high stiffness at the interface due to an imaging artefact (Fig. 4.3). All these problems can be overcome by using multiple CT datasets.

Three sets of CT data are needed—the intact bone, the bone after reaming and the bone after implantation (Fig. 4.4). The first set (the intact bone) will be used as the control model, which is the model without interfacial gaps. The second dataset (bone after reaming) will be used to identify interfacial gaps. This is done by creating a 3D bone model from the second dataset and the implanted femur from the third dataset. The two models were then aligned so that the model from the second dataset coincided with the model from the third dataset. The model of the implant, created separately in stereolithographic format, was then loaded into the software and aligned so that it coincided with the implant from the third CT dataset. Once the implant model was properly aligned, the third dataset and its model were removed leaving the implant model and the second CT dataset. The interfacial gaps can now be clearly identified (Fig. 4.5).

In order to create gaps at the appropriate place, the femoral cavity from the second dataset was segmented in the computer software to create a 3D model of the reamed canal. With the 3D model of the implant still aligned in its place, the

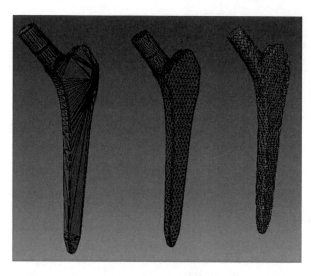

Fig. 4.2 The original CAD drawing from the manufacturer (*left*), the created surface-meshed model (*centre*) and model constructed from CT images (*right*)

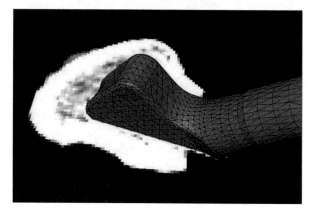

Fig. 4.3 The Alloclassic hip stem on a CT slice of an implanted femur, showing the artefacts at the interface (*strong white mark*) that translate into high stiffness at the interface using the Carter and Hayes density-stiffness relationship

location of the gaps can be identified based on the 3D model of the reamed canal that projected outside the surface of the implant (Fig. 4.6—*left*). These gaps were then turned into triangular surface elements by duplicating the elements on the surface of the implant and projecting them outwards (Fig. 4.6—*right*). The resulting implant model was then used as a cut-out for the bone model created from the first CT dataset. The cut-out procedure has been described in detail in Chap. 2. The resulting bone model therefore contained interfacial gaps at the appropriate places and perfect fit on the rest of the interface.

Another bone model to simulate perfect fit was created from the first CT dataset using the aligned implant without additional elements representing gaps. This bone

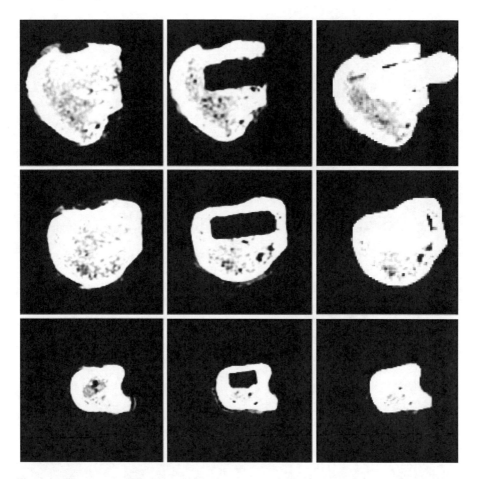

Fig. 4.4 The three sets of CT data—Intact bone (*left*), after reaming (*middle*) and after implantation (*right*)

model, together with the bone modelled with gaps, were then turned into solid tetrahedrals. The Alloclassic hip stem model was then merged into the two bone models separately, creating two separate FE models—one with interfacial gaps and another one with perfect interface fit.

Both models were then loaded in physiologic activities representing walking (Fisher's loadcase) and stair-climbing (Duda's loadcase). The material properties for the bone in both models were taken from the first CT dataset. All other FE parameters used in the models were the same as those used in Chap. 2.

4.1.2 Micromotion of the Stems

The results of micromotion are depicted in Fig. 4.7 for the Fisher's gait cycle (left) and Duda's stair-climbing (right). There were no significant differences between

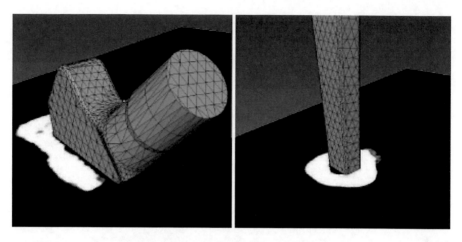

Fig. 4.5 Pictures showing the aligned Alloclassic and gaps at the interface

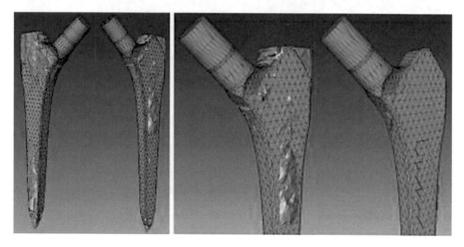

Fig. 4.6 Pictures showing interfacial gaps on the anterior and posterior side of the stem (*left*). The interfacial gaps turned into elements (*right*)

the models with interfacial gaps and the models with perfect fit in both physiological loadings. The overall distribution of micromotion is similar in each case, even though there are gaps at the interface, showing that these gaps did not endanger the stability of the prosthesis.

4.1.3 Finite Element Modelling: Second Method

Another comparison on the effect of gaps on primary stability was made using the ABG and the AML hip stems. The ABG is a proximal fixation design, where the

Fig. 4.7 Micromotion results using Fisher's loadcase (*left*) and Duda's loadcase (*right*). Each set contains the micromotion result for without gaps (*left*) and with gaps (*right*)

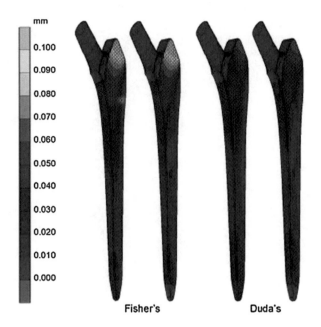

distal endosteal part is over-reamed to avoid cortical contact, a condition which may cause thigh pain [7]. Over-reaming the distal region also means that load will be transferred mostly through the proximal part of the implant. In this section, physical examination of interfacial gaps using CT datasets as described in the previous section could not be used due to the unavailability of specimens. Therefore, discussions with senior and experienced orthopaedic surgeons were carried out to determine the likelihood of interfacial gaps during femoral canal preparation. It was said that perfect fit in the proximal part of the ABG is crucial as this is the only area where loading is supposed to take place. However, the curved shape of the ABG meant that there are possibilities of gaps created during manual broaching in the lateral-proximal region. As a result, two FE models of the ABG were created, one with the lateral-proximal gap and another one with perfect proximal fit (Fig. 4.8—*middle and right*).

The ABG stem is anatomically shaped with distal endosteal over-reaming as standard procedure. However, there are other anatomically-shaped hip stems that do not use the same surgical procedure as the ABG. The PCA (Howmedica), the Profile (DePuy) and the Citation (Stryker) cementless hip stems, for example, are not distally over-reamed. Compared to the ABG, the PCA is the earlier anatomic hip stem. Complications from the use of PCA stems, such as thigh pain and adverse bone remodelling (distal bone hypertrophy and proximal bone atrophy) have led to the distally over-reamed anatomical ABG. It is unknown however if the gap in the distal region of the ABG system could compromise its primary stability. Based on this argument, another model was created with a perfect fit throughout the stem in order to compare the effect of distal over-reaming on anatomical hip stems (Fig. 4.8—*left*).

Fig. 4.8 Three cases of
the ABG stem analysed—
without interfacial gaps (*left*),
distal gap (*centre*) and distal
and lateral-proximal gap
(*right*)

There is one published report of a computer simulation study looking at the effect of totally removing the distal half of the ABG stem, thus allowing in theory complete proximal load transfer [8]. The authors concluded that the predicted bone remodelling pattern did not change much between the distally-cut ABG and the normal ABG. They also mentioned that reducing the stem length to the metaphyseal region only is not advantageous from a mechanical point of view. However, they did not carry out any stability simulation analyses.

The three ABG models were loaded with physiological walking and stair-climbing loads, with and without an interference fit of 100 μm. The bone model used was created from the VHP bone dataset and all other parameters remained the same as previous studies.

The other hip stem analysed in this section is the straight cylindrical AML. Contrary to the ABG, the AML is a distally-fixed cementless stem; fixation depends on a press-fit between the rough external surface of the implant and a similarly shaped femoral canal. Under-reaming surgical procedure is used to achieve tight fit with the cortical bone distally. However, achieving strong cortical support distally means that interfacial gaps could occur in the proximal region.

Three AML models were created—a model with a perfect interface fit as control, an AML with interfacial gaps in the proximal 1/3, and an AML with interfacial gaps in the proximal half (Fig. 4.9). The three models were loaded with physiological walking and stair-climbing loads with an interference fit of 100 μm. The VHP femur dataset was used as the bone model, and all other parameters remained the same as previous studies.

Fig. 4.9 Three cases of
the AML stem analysed—
without interfacial gap (*left*),
with gaps in the proximal 1/3
(*middle*) and with gaps in the
proximal 1/2 (*right*)

4.1.4 Micromotion of the Stems

Figures 4.10 and 4.11 show the contour plots of micromotion for the three ABG
models using Fisher's gait cycle and Duda's stair-climbing loads. For each loadcase (Fisher's/Duda's) two sets of results are displayed, without interference fit
(top pictures) and with an interference fit of 100 μm.

The general trend in all cases shows that micromotion increased as gaps were
introduced, first on the distal region and then in the distal + lateral proximal
region. There was a substantial increase in micromotion between the perfect fit
model and the other two ABG modelled with gaps. However, lateral-proximal
gaps that simulated surgical error during implantation did not appear to cause
any major changes in terms of magnitude or distribution of micromotion compared to the perfect proximal fit ABG during walking. In stair-climbing, on the
other hand, the existence of a lateral-proximal gap increased the micromotion
around this region significantly. There was also a considerable increase in micromotion between a perfect fit model and the distally over-reamed model in stair-
climbing compared to walking. The results also showed that the introduction of
an interference fit of 100 μm caused a considerable reduction in micromotion in
all cases.

Figure 4.12 shows the results for the AML hip stems under physiological
walking and stair-climbing loads. In both physiological loadings, the increase in
micromotion was found to be localised in the region around the gaps. Figure 4.13

Fig. 4.10 Micromotion results using Fisher's gait cycle forces for the ABG stem with distal gap (*middle*) and distal + lateral proximal gap (*right*) compared with a perfect fit model (*left*). Top pictures are results without interference fit, bottom pictures are results with an interference fit of 0.1 mm

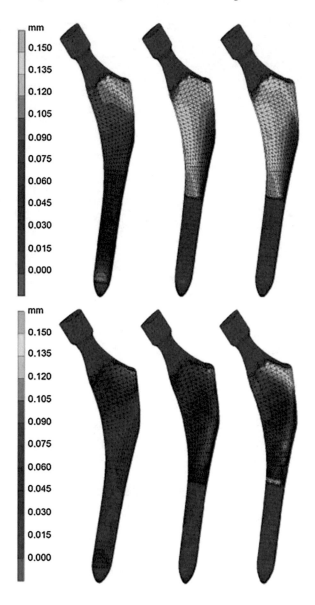

shows the surface area with micromotion in excess of the threshold limit of 50 μm for the two AML modelled with proximal gaps. The contour plots show that the surface area unfeasible for bone ingrowth in the AML modelled with gaps in the proximal ½ did not exceed beyond the region of the gaps, and the distribution and magnitudes of micromotion in the distal ½ of the stem were found to be unchanged. Similar results were obtained for the AML modelled with 1/3 proximal gaps.

Fig. 4.11 Micromotion results using Duda's stair-climbing forces for the ABG stem with distal gap (*middle*) and distal + lateral-proximal gap (*right*) compared with a perfect fit model (*left*). Top pictures are results without interference fit, bottom pictures are results with an interference fit of 0.1 mm

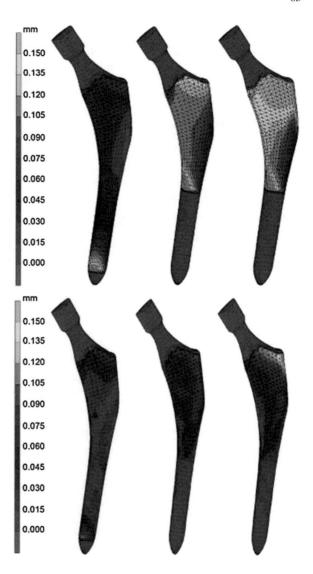

4.1.5 Biomechanical Evaluation of the Influence of Interfacial Gaps

Primary stability is important because it is an indicator of bone osseointegration and improved long-term fixation [9]. Poor initial stability might lead to bone resorption and encourage the formation of fibrous tissue at the interface [10]. It has been suggested that maximising the contact area of the implant surface with the host bone could improve the initial stability of the implant [1]. Since the shape of the prosthesis is exactly known, the opening of the femoral canal can be

Fig. 4.12 Micromotion
results using Duda's stair-
climbing forces (*top set*)
and Fisher's gait cycle
forces (*bottom set*) for the
AML stem with gaps in the
proximal 1/3 (*middle*) and
gaps in the proximal 1/2
(*right*), compared to a perfect
fit model (*left*)

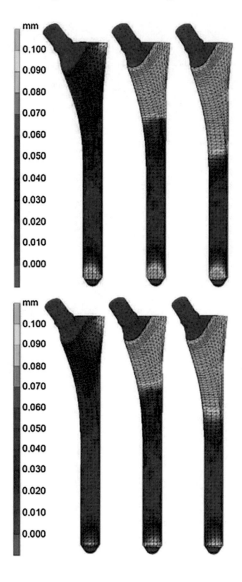

accurately milled with the use of surgical robots to create a precise cavity for the
implant used. If this hypothesis is true, the primary stability of an implant in a
robotically prepared cavity should be superior to that of a traditionally broached
cavity.

It has been reported that the use of surgical robotic systems improves the initial fit
as well as primary stability of the femoral components [11, 12]. This was attributed
to their findings that there were significantly more gaps between the stem and bone
in the hand-broached group compared to the robotic-reaming group. The stability of
implants in the traditionally-broached femora could be improved if compaction of the
bone bed and critical contact areas were achieved [13]. Thomsen et al. [14] reported

Fig. 4.13 Surface area
(*grey colour*) unfeasible for
bone ingrowth for the AML
modelled with 1/3 proximal
gap (*top set*) and ½ proximal
gap (*bottom set*). Results
using Fisher's walking forces
(*middle*) and Duda's stair-
climbing forces (*right*)

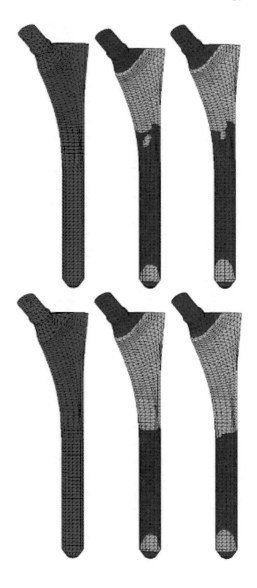

an experimental study comparing the effectiveness of robotic bone milling with traditional broaching as regards to primary rotational stability of 7 different cementless
stems. They found mixed results—some stem types were better in the robotic groups
whilst others were more stable in the hand-broached group. Robotic canal preparation did not improve stability in 4 of the 7 implants that they had tested; 2 of these
were anatomic in design. They have attributed this lack of stability on the gaps created by the robot because it needs more space for manoeuvering the drill head around
the curved shoulder of anatomic stems. Nogler et al. [15] have also found in their
experimental study that the robotically milled cavity did not show superior stability

for an anatomical stem compared to the manually broached one. However, they did not mention if their findings could be attributed to gaps created using both techniques during bone preparation.

This section looked at one of the possible surgical errors during implantation of hip prostheses—the creation of interfacial gaps—and its effects on primary stability. Three models were analysed—the rectangularly tapered Alloclassic, the anatomically shaped ABG and the straight cylindrical AML. The interfacial gaps in the Alloclassic were identified using a unique multiple CT scanning technique. For the ABG and the AML, only discussions with experienced surgeons could be conducted. Based on these discussions, the location of gaps for the ABG was assumed to be most probable in the lateral-proximal region and in the proximal area for the AML.

The existence of gaps due to surgical error did not affect the stability of the Alloclassic stem during common physiological activities of walking and stair-climbing. This is because the stability was achieved by wedging into the bone using the corners of the rectangular stem. The Alloclassic is designed to fit the femoral canal in the frontal plane but does not fill in the lateral plane. As long as press-fit is achieved by way of the four cortical corners along the stem, gaps at the non-critical areas such as the anterior and posterior surface of the stem do not endanger its stability.

For the ABG however, the stability seemed to be compromised in stair-climbing if there were gaps in the lateral-proximal region. Since the distal endosteal bone is over-reamed, the ABG relied mostly on a perfect fit of the ellipsoidal shape of the proximal part. This shape did not seem to be very stable under torsional load; slight surgical errors that introduce gaps could endanger the stability of the ABG hip stem. The Alloclassic, on the other hand, was better in sustaining torsional load because the four corners of the rectangularly-tapered design strongly anchored the stem to the endosteal bone.

This section also analysed the fixation of a distally fixed anatomical design such as the PCA and a distally over-reamed anatomical design such as the ABG. The PCA is the early cementless anatomic design where perfect surface to surface contact is used. Later anatomic designs such as the IPS (DePuy) and the ABG rely on the proximal fixation concept where gaps were intentionally created to avoid cortical contact distally. Our FE results showed that the perfect bone-implant contact design such as the one used in the PCA was more stable than the distally-over-reamed ABG. This is in accordance with the study by Dujardin et al. [16] who reported that well-fitted stems and a high percentage of canals fill reduced micromotion. This has also been analysed in the previous chapter when comparing between proximal versus distal fixation designs, where the distal fixation was found to be more stable than proximal fixation. However, if an interference fit of at least 100 μm could be achieved, a drastic improvement in terms of primary stability was predicted. It showed the importance of an interference fit especially in stems where there are gaps distally or otherwise.

Our FE results were in accordance with follow up studies reported in the literature for both the PCA and the ABG stems. A couple of follow-up studies of the PCA showed signs of adverse bone remodelling [17, 18]. Due to the distally fixed design of the PCA, bone hypertrophy was found distally and bone atrophy proximally. A 2-year follow-up study of the ABG showed that no mechanical complications such as loosening and migration can be found [19]. Distal hypertrophy, however, was identified but the extent was not as great as that found in the PCA.

The results of the AML hip stem showed that proximal gaps did not compromise the stability of the stem. In both physiological loadings, the surface area in excess of the threshold limit did not exceed beyond the region of the gaps, and the distribution and magnitude of micromotion in the distal ½ of the stem was found to be unchanged. The stability of the stem was therefore maintained by the distal half of the stem.

As discussed in the previous chapter, the AML femoral component performed well in hip arthroplasty. The results are well published, and good clinical results at 15 years could be obtained. The FE results in this study showed that one of the reasons to the success of the AML could be due to the stability of the stem in the presence of interfacial proximal gaps. As long as press-fit is achieved distally, the stem should be stable.

4.2 Undersizing and Malalignment

Another type of common surgical error is implant undersizing. Engh and Massin [20] for example, reported that out of 343 AML stems they had implanted, 42 % of them were found to be significantly undersized. They reported that these undersized stems statistically showed more late migration than correctly sized stems. Undersizing of stems occurred normally due to the cautiousness of surgeons who did not want to make the implant too tight in the canal as this would fracture or split the bone [21]. Undersized stems also have a tendency to shift into varus position—a malaligned situation [22].

Varus (medial tilt) or valgus (lateral tilt) malalignment is determined by measuring the angle made by the intersection of a line through the mid-stem of the prosthesis and a line through the midshaft of the femur as seen on the antero-posterior radiograph (Fig. 4.14) [23, 24].

Malalignment is not necessarily caused by shifting of the once-aligned undersized stem. It could also happen during bone preparation for implantation. If the reamer used is not properly aligned to the femoral canal, the tip of the reamer will touch the distal cortex sooner than it should. This will not only cause implant undersizing but also implant malalignment in a valgus or varus position.

The conventional way of planning for the correct implant size is by the use of template planning—a procedure where templates are placed over X-Ray images to help the surgeon decide the proper size for the patient. However, this procedure is not entirely reliable—the accuracy of templating improved as the level of training and experience increased, but even the most experienced surgeon could not achieve 100 % accuracy using the traditional 2D templating method. There are various published papers that evaluate planned-versus-achieved (PVA) accuracy or the repeatability of conventional template planning. A study carried out on the PCA cementless stems showed that the planned stem size was used in only 42 % of cases, the other

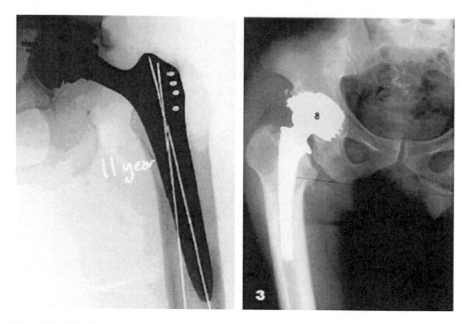

Fig. 4.14 AP radiographs showing varus malalignment of the Alloclassic hip stem (*left*—picture taken from Khalily and Whiteside 1998) and the Roy-Camille hip stem [25]

58 % had either undersized or oversized stems [26]. Similar results were reported by Carter et al. [27] who studied 74 cases of primary total hip arthroplasty with cementless femoral components, independently planned with templates by three surgeons. The templated size corresponded to the actual femoral component used in only 50 % of cases. The accuracy rose to 88–95 % if femoral components within one size below or above the templated size were included. Better templating in THA is possible through the use of three-dimensional computer-based planning [28]. The paper reported an accuracy of only 35 % for correct sizing using traditional templating compared to 52 % using computer-based planning. Their results also showed that there was a 25 % tendency to undersize when using conventional templating.

To the author's knowledge, there are no FE simulation studies analysing the effect of undersized or malaligned stems. This section of the chapter will therefore look at the effects of these surgical errors on primary hip stability. And since a malaligned implant is also most likely to be undersized, the analyses on malaligned stems will only be conducted on and compared to undersized stems.

4.2.1 Finite Element Modelling

Two types of hip stems were used in this study—the straight cylindrical AML and the rectangularly tapered Alloclassic. The bone model used in this study was the normal healthy bone taken from the VHP dataset. A suitable size that fits the

VHP bone model was identified to be size 5 for the Alloclassic, and size 135 for the AML was found to fit and fill the medullary canal. For the undersized stems of the Alloclassic, the original surface mesh of size 5 was scaled down into size 4 and size 3 using the actual size 5 as the template. For the undersized stems of the AML, since they are different in shape particularly in the proximal area, the two smaller sizes were created from the original CAD drawings obtained from the manufacturer. The two step sizes smaller than size 135 are the AML size 120 and size 105. These numbers represent the diameter of the stem's shaft—for example, size 135 represents a shaft diameter of 13.5 mm. These two undersized AMLs and the two undersized Alloclassics will be analysed and compared with their corresponding correct implant size that fits the femur model created from the VHP dataset.

For the malaligned implants, the smallest size from the above models of the Alloclassic (size 3) and the AML (size 105) were used. These implants were angulated in AMIRA software from their aligned position to represent ~5° varus malalignment (Fig. 4.15). There were problems however, involving the FE analyses with malaligned stems. The FE software (MARC.Mentat) could not solve the analyses when an interference fit was included. A solution could only be obtained if the interference fit was switched off. This could be because the distal region of the malaligned stems was not entirely in contact with solid bone. Bone marrow in the distal region of the canal caused problems in the FE software when an interference

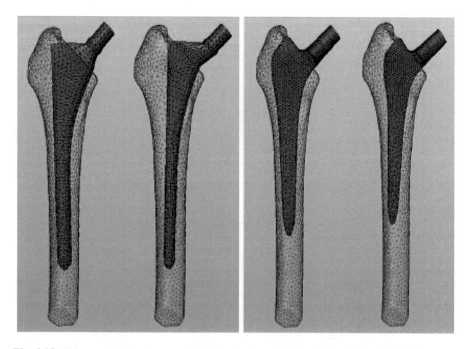

Fig. 4.15 Pictures showing the correct size implants and their undersized + malaligned counterparts (AML on *left*; Alloclassic on *right*)

fit was included. The interference fit was therefore excluded in all analyses in this section in order to make a proper comparison.

All other parameters were set to be the same as described in Chap. 2 and the models were loaded using both Duda's and Fisher's physiological loading conditions. The results for the undersized stems are shown first, followed by the results for the malaligned stems.

4.2.2 Biomechanical Assessment of Different Undersizing and Malalignment of the Stems

The results for the undersized stems (Figs. 4.16, 4.17, 4.18 and 4.19) showed that micromotion increased as the stem sizes were reduced for both implants and in both walking and stair-climbing. The results also showed that the undersized tapered design was better in terms of stability than an undersized cylindrical design. The AML stem with 2 sizes smaller than the correct size produced micromotion up to 10 times larger in Duda's loadcase. In Fisher's gait cycle, even one size smaller can cause a significant increase in micromotion. The result for the AML with two step sizes smaller failed to solve up to the maximum load, meaning that the stem would suffer gross migration. The Alloclassic, on the other hand, had a steady increase in micromotion as the sizes were reduced in both walking and stair-climbing.

For the malaligned stems (Figs. 4.20 and 4.21), there was a difference in primary stability between the tapered Alloclassic and the cylindrical AML. The

Fig. 4.16 Micromotion results for the Alloclassic size 5 (*left*), 4 (*centre*) and 3 (*right*) using Duda's stair-climbing loads

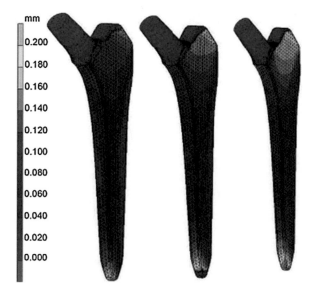

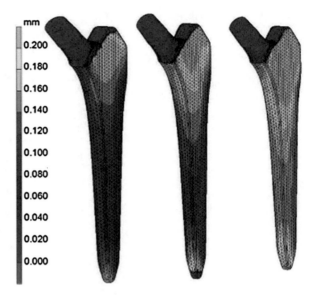

Fig. 4.17 Micromotion results for the Alloclassic size 5 (*left*), 4 (*centre*) and 3 (*right*) using Fisher's gait cycle loads

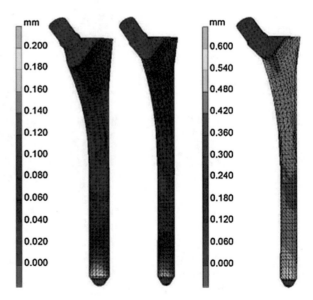

Fig. 4.18 Micromotion results for the AML stem size 135 (*left*), size 120 and with a different scale, size 105 (*right*) in stair-climbing

tapered design seemed to be better in achieving stability even when malaligned in a varus position. In contrast, a varus malaligned cylindrical design produced between 2 and 3 times more micromotion than a perfectly aligned implant.

The results showed that overall stability was compromised when an undersized implant was used, and the design of the implant played a crucial role in the extent of its instability. Two designs of hip stems were analysed in this study and to the author's knowledge, there have been no previous FE reports comparing the effect

Fig. 4.19 Micromotion results for the AML stem size 135 (*left*) and size 120 (*right*) in Fisher's gait cycle. Result for the AML size 105 was not included as the analysis failed before maximum load was achieved

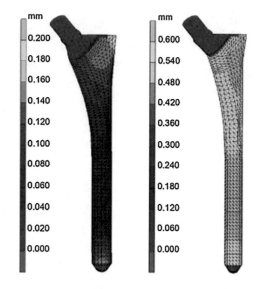

Fig. 4.20 Micromotion results for the Alloclassic size 3 in proper alignment and in varus malalignment. First set on the left using Duda's stair-climbing, and the second set on the right using Fisher's gait cycle

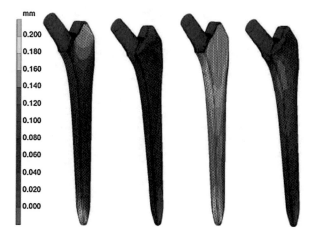

on primary stability between the two designs in undersized and malaligned position. However, there are several clinical follow-up studies reporting on the effects of undersized and malaligned stems.

Engh and Massin [20] reported that 42 % of 343 AML stems that they had implanted were significantly undersized. From the undersized group, there were 19 failures (15 unstable stems at 2 years, 2 late migrations at 4 years and 2 stem fractures). The difference in survivorship between canal-filling stems and undersized stems was also statistically significant (88 and 77 % respectively at 8 years). Among the 200 cases in which maximum stability due to correct sizing was achieved at the time of implantation, 94 % showed signs of early fixation by bone ingrowth and none was unstable at 2 years. In contrast, among the 143 cases in which the AML stem was undersized, only 60 % showed signs of early fixation by bone ingrowth.

Fig. 4.21 Micromotion results for the AML stem size 105 in proper alignment and in varus malalignment using Duda's stair-climbing loads (first two pictures). The last picture on the right showed micromotion results using Fisher's gait cycle for malaligned AML size 105. Analysis for properly aligned AML size 105 failed before maximum load was achieved

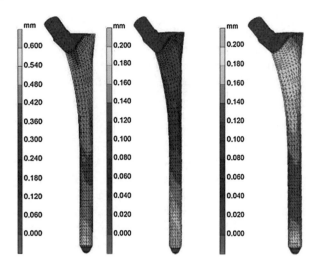

Another study looked at 52 hips which were followed for a minimum of 7 years after they had primary total hip arthroplasty with the AML hip system [21]. Three hips (6 %) which were undersized showed stable fibrous tissue ingrowth and another three undersized stems were found to be unstable. This can be compared with the 41 hips where the correct stem size was achieved, when 70 % had bone ingrowth and 30 % had stable fibrous tissue ingrowth. Another study [29] showed that undersized stems were more likely to have excessive subsidence (≥ 3 mm) than correctly sized stems. Three different stems were compared—the AML, the PCA and the LSF—and those stems that were undersized were found to be 19–25 % more likely to subside.

Giannikas et al. [7] also reported similar observations in their follow-up study of the ABG, a hip stem which has similar cylindrical feature of the AML. They also reported that revision surgery was performed on an undersized stem, because it caused persistent thigh pain due to significant subsidence of the prosthesis.

The FE results also showed that the micromotion was greater for the undersized implants than undersized + malaligned implants. The micromotion was smaller for the latter because the stem engaged the medial calcar proximally and the lateral endosteal cortex distally. This can be confirmed by a study on canine femoral prostheses, where the authors found that subsidence was significantly greater for the undersized implant in a neutral position than the undersized implant in varus malalignment [30]. They also reported that there was no significant difference in subsidence between the correctly sized implant in neutral position and the varus group.

For the Alloclassic hip stems, our FE analyses showed that this design is stable even if positioned in a varus alignment. This is in accordance with a follow-up study by Khalily and Lester [31]. No adverse effects were found on the Alloclassic hip stems with at least 5° of varus angulation at 5 years. They were all found to be clinically and radiographically stable. Another follow-up study also showed that

30 % of the Alloclassic hip stems had been implanted in slight varus position with no adverse clinical effect [32]. The report also showed that 16 % of the stems subsided more than 2 mm within the first 2 years after operation, but no progressive subsidence was detected after the second year. However, the author did not explain the cause of the subsidence. Based on our FE results, it could be that these stems were implanted undersized.

The FE results showed that undersized Alloclassic stems seemed to be better than undersized AML stems. This could be due to the differing geometrical design between the two stems. The taper in the Alloclassic provides a self-locking mechanism where a tighter fit is achieved under load, and the four corners of its rectangular shape provide the anchoring effect that is particularly effective in torsion. The undersized Alloclassic stem will therefore sink steadily under load until a stable position is achieved. This explanation is in agreement with a follow-up study on the Alloclassic stem reported by Delaunay et al. [33]. In their study, 4 stems were found to have subsided—3 stems between 2 and 5 mm and 1 stem subsided 5–10 mm within the first year. However, no progressive subsidence could be detected beyond this period–a condition which is usually termed late stabilisation. The AML stem, on the other hand, has a straight cylindrical shape which is much more unstable when implanted undersize because bone-anchoring cannot be achieved until it has sunk to the point where it rests on its proximal taper.

4.3 Pathological Condition of Bone

Another important parameter that could affect primary stability is the pathological condition of the bone. Skeletal diseases such as osteoporosis and osteoarthritis affect the bone's material properties. Osteoarthritic changes are mainly concentrated around the articulating surface of the joint where the cartilage covering the bone ends deteriorates due to age or injury. The properties of bone further away from the articulating joint may or may not change. Osteoporosis, on the other hand, is a skeletal disorder characterised by a significant loss of bone stock and structural deterioration of bone tissue. The bone becomes weak and fragile, and more likely to fracture under a sudden strain or fall.

Osteoarthritis is the most common cause of hip disease leading to primary hip replacement because it causes pain and severely reduced mobility. However, patients requiring hip arthroplasty also sometimes suffer from osteoporosis [29]. As osteoporotic bone is significantly weaker than normal healthy bone, it affects the decision made by the surgeon in terms of selection of suitable hip stems. It has been suggested that patients with osteoporosis would be better off having cemented hip stems to ensure strong primary fixation [34, 35]. However, Healy [1] reported that cementless stems were also reliable for elderly patients with poor bone stock.

Bone quality has been found to influence the extent of stress-shielding; severe bone loss was noted in poorer quality bones than in healthy ones [36, 37].

However, in terms of primary stability, it is unknown if weaker bone stock causes more micromotion and instability of the replaced hip. In general, stronger bone is preferred for stability, even though no conclusive evidence is available to support this hypothesis [1]. This is why cementless stems are still not widely used in patients with osteoporosis.

In Chap. 3, the effect of bone material properties on micromotion was analysed by reducing its stiffness values gradually to simulate weaker bone mass. This was done by reducing the constant value in the density-stiffness relationship proposed by Carter and Hayes (1977). In this section of the chapter, actual femoral CT datasets from patients suffering from osteoporosis and osteoarthritis will be used. The study will compare the interface micromotion and the stability of a cementless stem inside the two bone models (osteoarthritic and osteoporotic) with a healthy bone under physiological loadings. It is hoped that this study will give some qualitative data on the use of cementless stems on patients with weaker bone mass.

4.3.1 Finite Element Modelling

Data from patients about to undergo hip replacement surgery were obtained from the hospital. These data were originally being used by another researcher to study bone remodelling. All the patients were suffering from osteoarthritis of the hip joint, and the patients' femurs were CT scanned before the surgery. Out of 13 patients, only 1 femur was found to be osteoporotic as well. The Young Adult T-score of the patient, which is the World Health Organisation (WHO) criterion for osteoporosis, showed marked osteoporosis in all regions of the femur. The CT dataset of this patient as well as one other osteoarthritic bone were obtained, and segmented in AMIRA software. The CT dataset for the osteoporotic bone, however, was scanned separately for the proximal and middle parts with different voxel sizes. The models, therefore, had to be created separately because datasets from different voxel sizes cannot combine into a single dataset. The separate models were then joined together using MAGICS software as shown in Fig. 4.22. Material properties also need to be assigned separately for the proximal and middle part of the femur. A short computer code was written to help combine the large number of scattered material properties data from the two separate models. For the osteoarthritic bone, a single CT dataset was obtained. The construction of the bone model and the material properties assignment were therefore followed the procedure as described in Chap. 2.

Once the bone models had been created, the size of a suitable AML stem needed to be identified for each bone model. Four available sizes of the AML were manually positioned using AMIRA software, by orientating them until the stem was found to fit and fill the canal whilst maintaining the anteversion angle of the neck as described in Chap. 2. The best size to fit and fill the canal of the osteoporotic femur was found to be the largest size, AML-size150, and the suitable size for the osteoarthritic bone was found to be AML-size120. The stem size used for

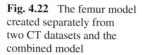

Fig. 4.22 The femur model created separately from two CT datasets and the combined model

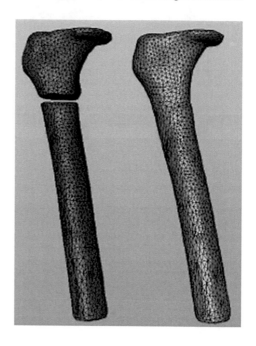

the normal bone was AML-size135, and the only difference between this and the other two sizes was the stem diameter; changed from 13.5 to 15 mm for the osteo-porotic and 12 mm for the osteoarthritic cases. The meshes of the two sizes of the AML stems were refined from their original 3D models and turned into solid tetrahedrals as described previously in Chap. 2. The models were then loaded as though in physiological stair-climbing using Duda's loadcase and walking using Fisher's loadcase. The micromotion results obtained were then compared with the micromotion for the implant in the normal bone from the VHP dataset (from Chap. 3).

4.3.2 Biomechanical Effect of Different Pathological Conditions of Bone

The first set of results show cut-through models of the three different bones, together with their corresponding stiffness distribution (Fig. 4.23). The contour plot showed that there was a large decrease in stiffness in the patient suffering osteoporosis compared to the healthy bone. There was also a marked reduction in thickness of the cortex on the anterior and posterior side of the osteoporotic femur. The contour plot of the osteoarthritic bone, however, showed slightly larger stiffness than the normal bone. Osteoarthritis is a joint disease that is concentrated on the area surrounding the joint. However, a comparative study made by Li and Aspden [38] showed that there appeared to be a small increase in mineralization

Fig. 4.23 The Young's modulus of bones from the normal VHP dataset (*left*), from a patient suffering from osteoporosis (*middle*) and osteoarthritis (*right*) (Pictures were scaled to fit)

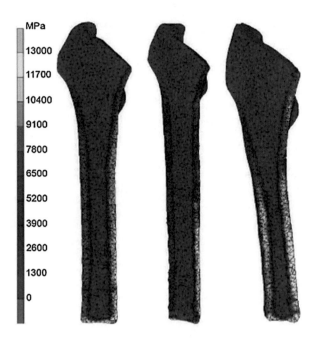

and about 70 % increase in the volume of trabecular bone in the osteoarthritic group compared to a normal group. They also found that increased apparent density of the OA trabecular bone resulted in a greater stiffness, yield strength and energy absorbed to yield.

The canal flare index was calculated for the three bone models using the technique proposed by Noble et al. [3]. The ratio of the canal diameter at two locations—20 mm above the lesser trochanter and the isthmus—was calculated from the antero-posterior view. The measurement was made from the FE software and the results are shown in Table 4.2. In their study, Noble et al. reported that canal flare indices of less than 3.0 described stovepipe canals (type C), 3.0–4.7 normal canals (type B), and 4.7–6.5, canals with a champagne-fluted appearance (type A). Table 4.2 shows that the normal bone used in this study is of type B, the osteoarthritic bone is of type A and the osteoporotic bone is of type C.

Figure 4.24 shows contour plots of micromotion under simulated physiological stair-climbing and walking. The figures show maximum micromotion for the osteoporotic model and minimum for the osteoarthritic bone model. The distribution of micromotion was, however, similar between all models, with larger micromotion observed at the proximal and distal of the stem. The difference in magnitude of micromotion between osteoporotic and normal bones ranged from a factor of 3 to a maximum of about 4. The maximum micromotion found for the osteoporotic model also reached 250 microns.

Table 4.3 shows the surface area with micromotion of 50 μm or larger for all three models. Overall, stair-climbing had more surface area beyond the threshold limit than walking. The osteoarthritic model was the most stable in both

Table 4.2 The canal flare index for the normal, osteoporotic and osteoarthritic bone models

	Normal	Osteoporotic	Osteoarthritic
Canal flare index	3.6	2.3	6.2

Fig. 4.24 The micromotion result from the normal VHP dataset (*left*), the osteoporotic bone (*middle*) and osteoarthritic bone (*right*) using Fishers's gait (*top*) and Duda's stair-climbing forces (bottom)

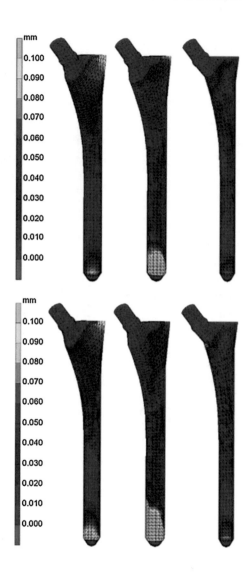

physiological loadings as it showed the least area. For the normal bone and the osteoporotic model, the surface area with micromotion larger than 50 μm depended on the type of activity. In walking, there was not much difference in the size of the area. In Duda's stair-climbing, however, there was almost twice as much surface area with micromotion ≥50 μm in the osteoporotic than in the normal bone.

Table 4.3 The amount of surface area more than 50 μm of micromotion. The total surface area for AML size 135 is 8,976 mm², AML size 150 is 9,902 mm² and AML size 120 is 7,858 mm²

		Area > 50 μm (mm²)	Percentage (%)
Fisher's gait	Normal	769	9
	Osteoporosis	754	8
	Osteoarthritis	44	1
Duda's stair-climbing	Normal	890	10
	Osteoporosis	1,776	18
	Osteoarthritis	78	1

In order to check for instability, interfacial bone loss is simulated in areas where micromotion exceeded the threshold micromotion limit. The new models with simulated interfacial bone loss were then loaded in Duda's stair-climbing mode as this was the loading with maximum impact on stability. The iterations continued until either a stable state micromotion is achieved or one of the two failure criteria, as described in Chap. 2, occurred. The results of the surface area unfeasible for bone ingrowth are shown in Table 4.4 and Fig. 4.25. The results show that the AML hip stem was unstable in the osteoporotic bone; the surface area unfeasible for bone ingrowth increased from 18 % in the first iteration to 78 % in the fourth iteration. The contour plots of micromotion for the osteoporotic model showing the increase in unfeasible surface area for bone ingrowth are shown in Fig. 4.26. The increase of the 'no growth' area in the normal and osteoarthritic bone, however, was minimal.

The results showed that bone quality does affect the stability of hip stems. In the case of the osteoarthritic model, stiffer cortical bone around the distal part of the stem and more cortical contact with the prosthesis increased the stability effectively. This is why the osteoarthritic model was found to be the most stable in both walking and stair-climbing. The osteoporotic bone, which had reduced bone stiffness and thinning of the cortex on the anterior and posterior sides showed significantly larger micromotion particularly in the distal area. Minimum contact with endosteal cortices and reduced cortical stiffness could be the reason for this. The FE analysis showed that patients who have had cementless hip replacement due to osteoarthritis alone should have fewer problems in terms of stability compared to patients suffering osteoporosis.

The results also showed that stair-climbing produced more micromotion and instability than walking. The effect was even more detrimental in the femur with weaker bone mass. When interfacial bone loss is simulated in the osteoporotic model, the surface area unfeasible for bone ingrowth increased from 18 to 78 %. In the normal and osteoarthritic models, on the other hand, the areas only increased slightly from 9 to 10 % and 1 to 2 % respectively. The findings suggest that patients who have had their hip replaced and who are also suffering from osteoporosis should take extra care especially during stair-climbing.

To the author's knowledge, there are no published reports comparing the primary stability of prostheses inside normal bone and bone with skeletal diseases.

Table 4.4 The increase in surface area with more than 50 μm of micromotion for the normal, osteoarthritic and osteoporotic models

	First iteration		Second iteration		Third iteration		Fourth iteration	
	Area > 50 μm (mm²)	(%)	Area > 50 μm (mm²)	(%)	Area > 50 μm (mm²)	(%)	Area > 50 μm (mm²)	(%)
Normal	769	9	890	10	933	10	–	–
OP	1,776	18	3,159	32	4,975	50	7,730	78
OA	78	1	154	2	188	2	–	–

Fig. 4.25 The reduction in surface area feasible for bone ingrowth

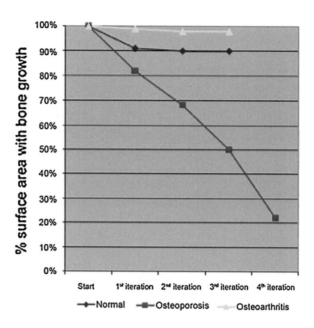

Fig. 4.26 Relative micromotion results for the osteoporotic model from the first iteration to the fourth iteration. The grey value is micromotion in excess of the threshold limit of 50 μm

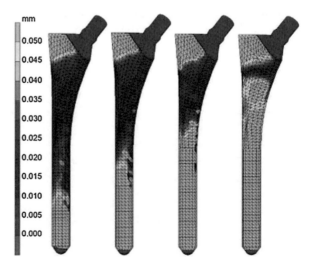

However, there are several reports showing that good bone quality is preferable to poor bone stock. Kim and Kim [21] for example showed that the AML stems used in younger patients had much lower frequency of loosening and revision rates compared to older patients. They have attributed this success to good bone quality, among other things. Krischak et al. [39] reported that femoral prosthesis loosening after seven years could be predicted by bone quality at the time of implantation—with loosening more likely if the bone quality was poor. A retrieval study by Maloney et al. [40] on various types of cemented and cementless stems showed that the extent of bone loss depended on the density of the bone; the less dense the bone was pre-operatively, the greater the extent of bone loss. This trend was also found by the study of Kerner et al. [36] when studying the AML stem.

The FE results in this section have been obtained from different AML sizes—the normal bone had size 135, the osteoporotic had size 150 and the osteoarthritic had size 120—due to the characteristics of the femoral canals. It could be that the FE results were affected not just by the material properties of the bone, but also by the different implant sizes used. However, in Chap. 3 the effect of weaker bone mass on micromotion was simulated by reducing the stiffness of the bone gradually from the Carter and Hayes relationship. The results from Chap. 3 were obtained using the VHP femoral dataset and size 135 of the AML, and they showed that micromotion increased as bone material properties get weaker. In conjunction with the results from Chap. 3, it is concluded that the stability of a cementless femoral prosthesis is compromised in weaker bone mass (such as in osteoporosis).

References

1. Healy WL (2002) Hip implant selection for total hip arthroplasty in elderly patients. Clin Orthop Relat Res 405:54–64
2. Mont MA, Hungerford DS (1997) Proximally coated ingrowth prostheses. A Rev Clin Orthop Relat Res 344:139–149
3. Noble PC, Alexander JW, Lindahl LJ, Yew DT, Granberry WM, Tullos HS (1988) The anatomic basis of femoral component design. Clin Orthop Relat Res 235:148–165
4. Paul HA, Bargar WL, Mittlestadt B, Musits B, Taylor RH, Kazanzides P, Zuhars J, Williamson B, Hanson W (1992) Development of a surgical robot for cementless total hip arthroplasty. Clin Orthop Relat Res 285:57–66
5. Testi D, Simeoni M, Zannoni C, Viceconti M (2004) Validation of two algorithms to evaluate the interface between bone and orthopaedic implants. Comput Meth Prog Bio 74(2):143–150
6. Pazzaglia UE, Brossa F, Zatti G, Chiesa R, Andrini L (1998) The relevance of hydroxyapatite and spongious titanium coatings in fixation of cementless stems. An experimental comparative study in rat femur employing histological and microangiographic techniques. Arch Orthop Trauma Surg 117(4–5):279–285
7. Giannikas KA, Din R, Sadiq S, Dunningham TH (2002) Medium-term results of the ABG total hip arthroplasty in young patients. J Arthroplasty 17(2):184–188
8. van Rietbergen B, Huiskes R (2001) Load transfer and stress shielding of the hydroxyapatite-ABG hip: A study of stem length and proximal fixation. J Arthroplasty 16 (8, Supplement 1):55–63
9. Pilliar RM, Lee JM, Maniatopoulos C (1986) Observations on the effect of movement on bone ingrowth into porous-surfaced implants. Clin Orthop Relat Res 208:108–113

10. Soballe K, Overgaard S, Hansen ES, Brokstedt-Rasmussen H, Lind M, Bunger C (1999) A review of ceramic coatings for implant fixation. J Long Term Eff Med Implants 9(1–2):131–151
11. Bargar WL, Bauer A, Borner M (1998) Primary and revision total hip replacement using the Robodoc system. Clin Orthop Relat Res 354:82–91
12. Paravic V, Noble PC, McCarthy JC (1999)The impact of robotic surgery on the fit of cementless femoral prostheses, 45th edn. Orthopaedic Research Society, Anaheim
13. Alexander JW, Kamaric E, Noble PC, McCarthy JC (1999) The effect of robotic machining on the micromotion of cementless femoral stems, 45th edn. Anaheim, Orthopaedic Research Society
14. Thomsen MN, Breusch SJ, Aldinger PR, Gortz W, Lahmer A, Honl M, Birke A, Nagerl H (2002) Robotically-milled bone cavities: a comparison with hand-broaching in different types of cementless hip stems. Acta Orthop Scand 73(4):379–385
15. Nogler M, Polikeit A, Wimmer C, Bruckner A, Ferguson SJ, Krismer M (2004) Primary stability of a robodoc implanted anatomical stem versus manual implantation. Clin Biomech (Bristol, Avon) 19 (2):123–129
16. Dujardin FH, Mollard R, Toupin JM, Coblentz A, Thomine JM (1996) Micromotion, fit, and fill of custom made femoral stems designed with an automated process. Clin Orthop Relat Res 325:276–289
17. Jacobsson SA, Djerf K, Gillquist J, Hammerby S, Ivarsson I (1993) A prospective comparison of Butel and PCA hip arthroplasty. J Bone Joint Surg Br 75(4):624–629
18. Jansson V, Refior HJ (1992) Clinical results and radiologic findings after cementless implantation of PCA stems in total hip replacement. Arch Orthop Trauma Surg 111(6):305–308
19. Tonino AJ, Romanini L, Rossi P, Borroni M, Greco F, Garcia-Araujo C, Garcia-Dihinx L, Murcia-Mazon A, Hein W, Anderson J (1995) Hydroxyapatite-coated hip prostheses. Early results from an international study. Clin Orthop Relat Res 312:211–225
20. Engh CA, Massin P (1989) Cementless total hip arthroplasty using the anatomic medullary locking stem. Results using a survivorship analysis. Clin Orthop Relat Res 249:141–158
21. Kim YH, Kim VE (1994) Cementless porous-coated anatomic medullary locking total hip prostheses. J Arthroplasty 9(3):243–252
22. Ries MD, Lynch F, Jenkins P, Mick C, Richman J (1996) Varus migration of PCA stems. Orthopedics 19 (7):581–585, discussion 585–586
23. Khalily C, Whiteside LA (1998) Predictive value of early radiographic findings in cementless total hip arthroplasty femoral components: An 8- to 12-year follow-up. J Arthroplasty 13(7):768–773
24. RÃkkum M, Reigstad A (1999) Total hip replacement with an entirely hydroxyapatite-coated prosthesis: 5 years' follow-up of 94 consecutive hips. J Arthroplasty 14(6):689–700
25. Chen CH, Shih CH, Lin CC, Cheng CK (1998) Cementless Roy-Camille femoral component. Arch Orthop Trauma Surg 118(1–2):85–88
26. Knight JL, Atwater RD (1992) Preoperative planning for total hip arthroplasty. Quantitating its utility and precision. J Arthroplasty 7(Suppl):403–409
27. Carter LW, Stovall DO, Young TR (1995) Determination of accuracy of preoperative templating of noncemented femoral prostheses. J Arthroplasty 10(4):507–513
28. Viceconti M, Lattanzi R, Antonietti B, Paderni S, Olmi R, Sudanese A, Toni A (2003) CT-based surgical planning software improves the accuracy of total hip replacement preoperative planning. Med Eng Phys 25(5):371–377
29. Haddad RJ, Cook SD, Brinker MR (1990) A comparison of 3 varieties of noncemented porous-coated hip replacement. J Bone Joint Surg Br 72(1):2–8
30. Pernell RT, Gross RS, Milton JL, Montgomery RD, Wenzel JG, Savory CG, Aberman HM (1994) Femoral strain distribution and subsidence after physiological loading of a cementless canine femoral prosthesis: The effects of implant orientation, canal fill, and implant fit. Vet Surg 23(6):503–518
31. Khalily C, Lester DK (2002) Results of a tapered cementless femoral stem implanted in varus. J Arthroplasty 17(4):463–466

32. Delaunay C, Cazeau C, Kapandji AI (1998) Cementless primary total hip replacement—Four to eight year results with the Zweymuller-Alloclassic (R) prosthesis. Int Orthop 22:1–5

33. Delaunay C, Fo Bonnomet, North J, Jobard D, Cazeau C, Kempf J-Fo (2001) Grit-blasted titanium femoral stem in cementless primary total hip arthroplasty: A 5- to 10-year multi-center study. J Arthroplasty 16(1):47–54

34. Dorr LD, Wan Z, Gruen T (1997) Functional results in total hip replacement in patients 65 years and older. Clin Orthop Relat Res 336:143–151

35. Haber D, Goodman SB (1998) Total hip arthroplasty in juvenile chronic arthritis: a consecutive series. J Arthroplasty 13(3):259–265

36. Kerner J, Huiskes R, van Lenthe GH, Weinans H, van Rietbergen B, Engh CA, Amis AA (1999) Correlation between pre-operative periprosthetic bone density and post-operative bone loss in THA can be explained by strain-adaptive remodelling. J Biomech 32(7):695–703

37. Ohsawa S, Fukuda K, Matsushita S, Mori S, Norimatsu H, Ueno R (1998) Middle-term results of anatomic medullary locking total hip arthroplasty. Arch Orthop Trauma Surg 118(1–2):14–20

38. Li B, Aspden RM (1997) Material properties of bone from the femoral neck and calcar femorale of patients with osteoporosis or osteoarthritis. Osteoporos Int 7(5):450–456

39. Krischak GD, Wachter NJ, Zabel T, Suger G, Beck A, Kinzl L, Claes LE, Augat P (2003) Influence of preoperative mechanical bone quality and bone mineral density on aseptic loosening of total hip arthroplasty after seven years. Clin Biomech (Bristol, Avon) 18 (10):916–923

40. Maloney WJ, Woolson ST (1996) Increasing incidence of femoral osteolysis in association with uncemented Harris-Galante total hip arthroplasty. J Arthroplasty 11(2):130–134 A follow-up report

Summary

Cementless hip stems are gaining popularity in hip joint arthroplasty but the issues related to their stability are a major concern. This monograph uses Finite Element Analyses to study various factors that affect the stability of cementless hip stems. A micromotion algorithm was written in Compaq Visual Fortran to calculate and display interface micromotion—a parameter that describes the stability of hip stems. The computer code was checked for accuracy using published data of a simplified cylindrical bone-implant model. The model was also used to study the effect of location for measurement of relative micromotion at the interface.

In this monograph, two types of micromotion results were presented. The first was simply the contour plots of implant micromotion relative to the bone. However, this method alone could not answer the stability of the stem. A novel technique was used, where bone loss was simulated at the interface where micromotion exceeded the threshold limit for bone ingrowth. The analysis was repeated until either a stable-state micromotion was achieved or the implant failed – either by exceeding the interfacial shear strength or the implant surface was encapsulated with the threshold micromotion limit, i.e. encapsulated with fibrous tissue. This technique was used to analyse various factors that could affect the stability of cementless hip stems.

The major part of this monograph looked at the effect of hip stem designs on interface micromotion and stability. The stems were categorised into three basic types based on the overall geometry—straight cylindrical, tapered, and anatomical designs. All three models representing the three groups had similar magnitude and distribution of micromotion. When interfacial bone loss was simulated, all of them were found to be stable, with the anatomical design the most stable in both simulated walking and stair climbing.

Three elastic moduli representing cobalt chromium, titanium alloy and a composite material were also analysed under physiological loading. The contour plots of micromotion confirmed the results from other published reports—the more flexible the implant, the larger the micromotion. However, with the proposed technique of measuring instability, all three implants were predicted to be stable. The composite stem was still the worst, with predicted fibrous tissue formation

M. R. Abdul Kadir, *Computational Biomechanics of the Hip Joint*, SpringerBriefs in
Computational Mechanics, DOI: 10.1007/978-3-642-38777-7,
© The Author(s) 2014

covering the proximal half of the stem. However, the stem should still be stable if tight fit was achieved distally.

Four hip stems with a decreasing stem length were also modelled, with the shortest one having minimal cortical bone contact. Micromotion increased as the length was shortened, with the shortest stem having micromotion in excess of the threshold limit covering almost the whole surface. When bone loss was simulated, the first three stems were found to be stable. As long as cortical contact was achieved with press-fit, stability was not compromised.

The comparative micromotion results also showed that the distribution of micromotion was different between the proximal and distal designs. Less micromotion was found in the lateral area of the proximal design, which could be attributed to the effectiveness of the lateral flare feature. However, more micromotion was found in the medial area of the proximal design than the distal design. When bone loss was simulated, the proximal design was found to be unstable compared to the distal design. The total loss of fixation was due to the loss of bone contact in the medial region.

The last chapter of the monograph looked into the effect of surgical and pathological parameters on micromotion. During bone preparation, broaching and reaming could create gaps at the interface. A technique of identifying the location of interfacial gaps using multiple CT datasets was proposed, and implemented on the Alloclassic hip stem. The gaps created during bone preparation did not affect the stability of the stem as long as press-fit was achieved by way of the four corners along the stem. Apart from the Alloclassic hip stem, two other hip stem designs were also analysed—the straight cylindrical AML and the anatomical ABG. The possible locations of interfacial gaps for the AML and the ABG hip systems were estimated. The interface micromotion was larger in the hip stems with distal bone over-reaming compared to a design with a perfect bone-implant contact. Achieving interference fit in a proximally-fixed cementless hip stem is also important as it reduces the effect of interfacial gaps on stability. The stability of a distally fixed straight cylindrical hip stem was not compromised by the presence of proximal interfacial gaps as long as interference fit was achieved distally.

The effect of hip stems malalignment and undersizing due to surgical error was also carried out. Two types of hip stems were analysed—the AML and the Alloclassic. Undersizing should always be prevented as it increased micromotion and thus increasing implant instability. The design of the implant determined the magnitude of instability - the straight cylindrical design was found to be more susceptible to instability when undersized components were used compared to the tapered design. For hip stems in varus malalignment, both designs were relatively more stable than the undersized with normally aligned stems, because the varus stems rested on the medial calcar proximally and the lateral cortex distally. In varus malalignment, the tapered design was more stable than the straight cylindrical design.

For the effect of pathological condition on stability, two femoral models were constructed from two CT datasets of patients suffering from osteoarthritis and osteoporosis respectively. The canal flare indices for the two bones, as well as the normal healthy bone from the VHP dataset were calculated. The normal and osteo-arthritic bones had minor effect on micromotion and were found to be stable when bone loss was simulated. Osteoporotic bone, however, had the largest area of inter-face micromotion in excess of the threshold limit especially during stairclimbing, and was found to be unstable when bone loss was simulated.

Index

M. R. Abdul Kadir, *Computational Biomechanics of the Hip Joint*, SpringerBriefs in
Computational Mechanics, DOI: 10.1007/978-3-642-38777-7,
© The Author(s) 2014